PCR

Second Edition

C.R. Newton and A. Graham

ZENECA Pharmaceuticals, Mereside, Alderley Park, Macclesfield, Cheshire SK10 4TG, UK

βIOS
SCIENTIFIC
PUBLISHERS

© BIOS Scientific Publishers Limited, 1994, 1997

First published 1994 (ISBN 1 872748 82 1)
Second Edition 1997 (ISBN 1 85996 011 1)

A CIP catalogue record for this book is available from the British Library.

ISBN 1 85996 011 1 second edition

BIOS Scientific Publishers Ltd
9 Newtec Place, Magdalen Road, Oxford OX4 1RE, UK
Tel. +44 (0) 1865 726286. Fax. +44 (0) 1865 246823
World Wide Web home page: http://www.Bookshop.co.uk/BIOS/

DISTRIBUTORS

Australia and New Zealand
    DA Information Services
    648 Whitehorse Road, Mitcham
    Victoria 3132

India
    Viva Books Private Limited
    4325/3 Ansari Road
    Daryaganj
    New Delhi 110002

Published in the United States of America, its dependent territories and Canada by Springer-Verlag New York Inc. in association with BIOS Scientific Publishers Ltd. Copies may be obtained from:

Springer-Verlag New York Inc.
175 Fifth Avenue, New York
NY 10010-7858, USA

Typeset by Chandos Electronic Publishing, Stanton Harcourt, UK.
Printed by Information Press, Eynsham, Oxon, UK.

PCR

Second Edition

The INTRODUCTION TO BIOTECHNIQUES series

Editor:

D. Billington School of Biomolecular Sciences, Liverpool John Moores University, Byrom Street, Liverpool L3 3AF

CENTRIFUGATION
RADIOISOTOPES
LIGHT MICROSCOPY
ANIMAL CELL CULTURE
GEL ELECTROPHORESIS: PROTEINS
MICROBIAL CULTURE
ANTIBODY TECHNOLOGY
GENE TECHNOLOGY
LIPID ANALYSIS
GEL ELECTROPHORESIS: NUCLEIC ACIDS
LIGHT SPECTROSCOPY
PCR, SECOND EDITION

Forthcoming titles

PLANT CELL CULTURE
MEMBRANE ANALYSIS

Contents

PART 2: TECHNIQUES AND APPLICATIONS

4. Cloning and Modification of PCR Products   47

5. Isolation and Construction of DNA Clones   63

Abbreviations

A	adenine
AFLPs	amplified fragment length polymorphisms
AMDMIP	aminomethyl-4, 5′-dimethylisopsoralen
AMV	avian myeloblastosis virus
ARMS	amplification refractory mutation system
ASO	allele-specific oligonucleotide
ATP	adenosine triphosphate
BAC	bacterial artificial chromosome
bp	base pair
C	cytosine
CCD	charge coupled device
CCM	chemical cleavage of mismatches
cDNA	complementary DNA
CF	cystic fibrosis
CFTR	cystic fibrosis transmembrane conductance regulator
CML	chronic myelogenous leukemia
CM-PCR	chromosome microdissection PCR
COP	competitive oligonucleotide priming
dATP	deoxyadenosine triphosphate
dCTP	deoxycytidine triphosphate
M5dCTP	5-methyldeoxycytosine triphosphate
ddNTP	dideoxynucleoside triphosphate
DD-PCR	differential display PCR
DDRT-PCR	differential display reverse transcription PCR
DEVIATS	double-ended vectorette incorporating alternative transcription sites
DGGE	denaturing gradient gel electrophoresis
dGTP	deoxyguanosine triphosphate
DMD	Duchenne muscular dystrophy
DMSO	dimethylsulfoxide
DNA	deoxyribonucleic acid
dNMP	deoxynucleoside monophosphate
dNTP	deoxynucleoside triphosphate
dsDNA	double-stranded DNA
dTMP	deoxythymidine monophosphate
dTTP	deoxythymidine triphosphate
dUMP	deoxyuridine monophosphate
ECL	electrochemiluminescence
EC-PCR	expression cassette-PCR
EDTA	ethylenediaminetetraacetic acid
ELISA	enzyme-linked immunosorbent assay
E-PCR	expression-PCR
EST	expressed sequence tag
EtBr	ethidium bromide
FAM	6-carboxyfluorescein
G	guanine
GAWTS	gene amplification with transcript sequencing
GSP	gene-specific primer
HBV	hepatitis B virus

HEX	4,7,2′,4′,5′,7′-hexachloro-6-carboxyfluorescein
HIV	human immunodeficiency virus
HLA	human leukocyte antigen
HPLC	high-pressure liquid chromatography
HPV	human papilloma virus
HTLV	human T-cell lymphotrophic virus
kbp	kilobase pair
LA-PCR	ligation-anchored PCR
LD-PCR	long-distance PCR
LIC	ligation-independent cloning
MAPPing	message amplification phenotyping
MMLV	Moloney murine leukemia virus
8-MOP	8-methoxypsoralen
mRNA	messenger RNA
mtDNA	mitochondrial DNA
nt	nucleotides
OD	optical density
PAC	P1-derived artificial chromosome
PAMSA	PCR amplification of multiple specific alleles
PCR	polymerase chain reaction
PEG	polyethyleneglycol
PEST	primer extension sequence test
RACE	rapid amplification of cDNA ends
RAPD	random amplified polymorphic DNA
RAWIT	RNA amplification with *in vitro* translation
RAWTS	RNA amplification with transcript sequencing
RBS	ribosome-binding site
RFLP	restriction fragment length polymorphism
RLM-RACE	RNA ligase-mediated RACE
RLPCR	reverse ligation-mediated PCR
RNA	ribonucleic acid
RNase	ribonuclease
rRNA	ribosomal RNA
RT	reverse transcriptase
SDS	sodium dodecyl sulfate
SOE	splicing by overlap extension
SPA	scintillation proximity assay
SSCP	single-strand conformation polymorphism
STS	sequence tagged site
T	thymine
TBR	Tris(2,2′-bipyridine)ruthenium (II) chelate
TD-PCR	touchdown PCR
TET	4,7,2′,7′-tetrachloro-6-carboxyfluorescein
T_m	melting temperature
TMAC	tetramethylammonium chloride
U	uracil
UNG	uracil N-glycosylase
USE	unique site elimination
UV	ultraviolet
VNTRs	variable number tandem repeats
YAC	yeast artificial chromosome

Preface

In the 10 years since its introduction the polymerase chain reaction (PCR) has become a basic and essential tool in both research and analytical laboratories. Indeed, a computer search of a bioscience literature database performed during the preparation of this book revealed that in 1985 and 1986 there were just 38 and 61 PCR citations, respectively. The number of references has increased annually in an exponential manner akin to the product in a PCR. At the time of writing this, the second edition, there were more than 6×10^4 PCR references.

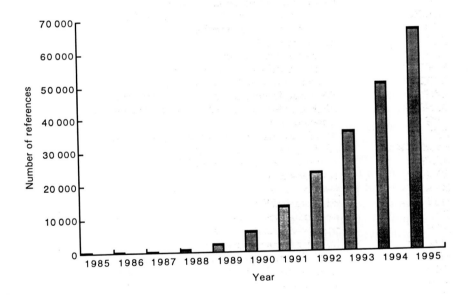

Cumulative citations of the polymerase chain reaction during its first decade.

The fundamental PCR has been adapted for a range of specialist applications ranging from the characterization of genes, their cloning and expression to DNA diagnostics where PCR is used in the detection of pathogens, identifying mutations responsible for inherited diseases and DNA fingerprinting for medical and forensic reasons. By virtue of the speed, sensitivity, specificity and inherent simplicity of the PCR, it has become the method of choice for the above applications in most laboratories. Although several texts exist that are devoted solely to the PCR, these are fundamentally dedicated to the practical aspects of the technique, containing detailed protocols for each specialist PCR application. This book, although intended as a practical guide, also concentrates on the fundamentals and capabilities of the reaction and explains the attributes of the reaction in the context of each application.

This book is aimed at those new to PCR but we hope that it will also prove a useful text to those more experienced in applying the PCR and that the information provided is useful as further guidance on familiar techniques and as a source of reference for less commonly used applications. Part 1 describes the basic PCR in detail and continues to examine the equipment and thermophilic enzymes available to carry out the PCR. Part 1 then proceeds to give the essential information needed to perform successful and specific PCRs and examines the role of primers in the reaction; in this context, the design and modification of primers is also considered.

Part 2 details each of the specialist techniques and applications for which the PCR is used. The various methods for cloning, modifying, joining and mutating PCR products are described first. Next, the methods that allow these to be sequenced are covered. Part 2 then goes on to examine how PCR can be used to determine novel DNA sequence information and how this can be applied to genome analysis. This is followed by fingerprinting of genomes and ways PCR can be used to examine genetic variation. Methods of characterizing both unknown and known mutations follow. Finally, Part 2 concludes with two chapters covering detecting and characterizing pathogens and the developments that have allowed the PCR technique to be employed in a quantitative mode.

Clive Newton
Alex Graham

Acknowledgements

We would like to acknowledge gratefully the generosity of the following friends, colleagues and organizations for contributing figures and tables appearing in this book: P.E. Applied Biosystems (*Table 1.1*); Dr Keith Edwards of Integrated Approach to Crop Research, Long Ashton University of Bristol (*Figures 9.2* and *9.3*); and Dr Mike Litt of The Oregon Health Sciences University (*Figure 9.4*); Dr Nada Ghanem of the Institut National de la Santé et de la Recherche Médicale (*Figures 10.3* and *10.5*); Dr Ann Harris and Ms S. Shackleton of the Institute of Molecular Medicine, John Radcliffe Hospital (*Figure 10.7*); Dr Roland Roberts of the Paediatric Research Unit, Guy's Hospital (*Figure 10.8*); Ms Geraldine Malone of the Paediatric Genetics Unit, Royal Manchester Children's Hospital (*Figure 11.3*); Dr Jo Whittaker and Ms Rachel Butler of The Institute of Child Health, Alder Hey Children's Hospital (*Figure 11.4*); Dr Steve Little and Ms Nancy Robertson of Zeneca Diagnostics (*Figure 11.5*).

While every care has been taken to ensure that the experimental details discussed in this book meet all relevant safety requirements, the authors accept no liability for any loss or injury howsoever caused.

The reader is hereby notified that the purchase of this book does not convey any license or authorization to practice PCR under any patents assigned to or applied for by Hoffmann La Roche Inc. In addition, we recommend the purchase of reagents from authorized suppliers only. At the time of going to press these comprised: A.B. Fermentas, Lithuania; Advanced Bioenzymes, UK; Amersham Bioscience, England; Biotech International Ltd, Australia; Boehringer Mannheim America, USA; CLONTECH Laboratories, USA; DNAmp, UK; Enzyme Technologies, UK; Life Technologies Inc. USA; Nippon Gene Co. Ltd, Japan; Pharmacia Biotech, Sweden; Qiagen, Germany; Stratagene, USA; Takara Shuzo Co. Ltd, Japan and Toyobo, Japan.

1 What is PCR?

The polymerase chain reaction (PCR) is an *in vitro* technique which allows the amplification of a specific deoxyribonucleic acid (DNA) region that lies between two regions of known DNA sequence. In defining PCR we must first consider the DNA molecule. In its native state, DNA exists as a double helix. This helix comprises two single

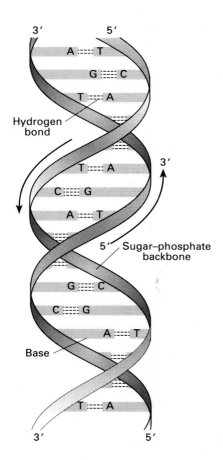

FIGURE 1.1: *The DNA double helix. Figure reproduced from Williams et al. (1993)* Genetic Engineering, *BIOS Scientific Publishers Ltd.*

strands of DNA running antiparallel to each other and held together noncovalently by hydrogen bonds. The hydrogen bonds form between the complementary bases adenine (A) with thymine (T) and guanine (G) with cytosine (C) (see *Figure 1.1*). The structures and ring numbering of the bases is shown in *Figure 1.2*. The bases are each attached to a sugar molecule, deoxyribose, and each sugar molecule is joined to the adjacent sugar molecule via a phosphate group. One 'unit' of DNA comprises a phosphate group, a sugar and a base and is known as a nucleotide; a sugar and base alone is known as a nucleoside. The structure of a four-nucleotide segment of DNA is shown in *Figure 1.3*. Note the numbering of the carbon atoms of the sugar part of the molecule. To distinguish between the base and the sugar numbering, the numbers for the sugar carbons each have a 'prime' (e.g. 5' and 3'). It is the 5' and 3' carbons of adjacent sugars that are linked via the phosphate groups, hence single-stranded DNA and the related molecule, ribonucleic acid (RNA) will have a 5' end and a 3' end. The 5' end of one strand of double-stranded DNA is complementary to the 3' end of the other strand. RNA has the base uracil (U) instead of T, and has an additional hydroxyl group on its sugar moieties (ribose) at the 2' position. In RNA, U can base-pair with A.

PCR amplification of DNA is achieved by using oligonucleotide primers, also known as amplimers. These are short, single-stranded DNA molecules which are complementary to the ends of a defined sequence of DNA template. The primers are extended on single-stranded denatured DNA (template) by a DNA polymerase, in the presence of deoxynucleoside triphosphates (dNTPs) under suitable reaction conditions. This results in the synthesis of new DNA strands

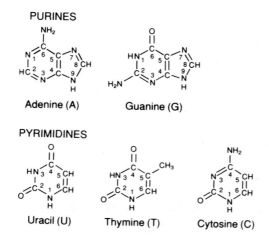

FIGURE 1.2: *The structure and ring numbering of the bases. Attachment of the bases is via nitrogen 1 of the pyrimidine bases and via nitrogen 9 of the purine bases.*

complementary to the template strands. These strands exist at this stage as double-stranded DNA molecules. Strand synthesis can be repeated by heat denaturation of the double-stranded DNA, annealing of primers by cooling the mixture and primer extension by DNA polymerase at a temperature suitable for the enzyme reaction. Each repetition of strand synthesis comprises a cycle of amplification. Each new DNA strand synthesized becomes a template for any further cycle of amplification and so the amplified target DNA sequence is selectively amplified cycle after cycle. *Figure 1.4* shows the first few cycles of PCR. The first extension products result from

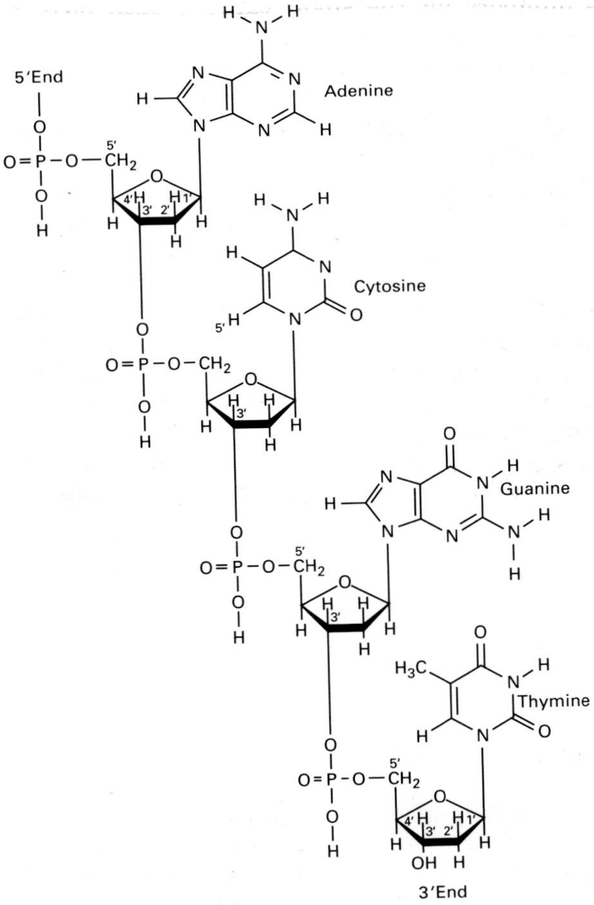

FIGURE 1.3: A four-nucleotide sequence of single-stranded DNA. Figure redrawn from Williams et al. (1993) Genetic Engineering, BIOS Scientific Publishers Ltd.

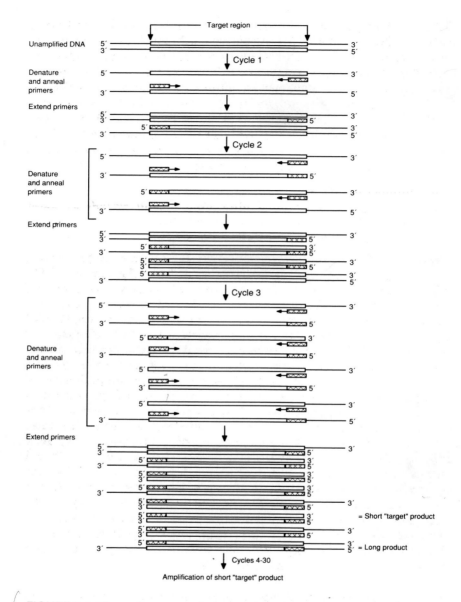

Amplification of short "target" product

FIGURE 1.4: *The polymerase chain reaction. PCR is a cycling process; with each cycle the number of DNA targets doubles. The strands in the targeted DNA are separated by thermal denaturation and then cooled to allow primers to anneal specifically to the target region. DNA polymerase is then used to extend the primers in the presence of the four dNTPs and suitable buffer. In this way duplicates of the original target region are produced and this 'cycle' is normally repeated for 20–40 cycles. The short 'target' products, which increase exponentially after the fourth cycle, and the long products, which increase linearly, are shown.*

DNA synthesis on the original template and these do not have a distinct length as the DNA polymerase will continue to synthesize new DNA until it either stops or is interrupted by the start of the next cycle. The second cycle extension products are also of indeterminate length; however, at the third cycle, fragments of 'target' sequence are synthesized which are of defined length corresponding to the positions of the primers on the original template. From the fourth cycle onwards the target sequence is amplified exponentially. Thus, amplification, as the final number of copies of the target sequence, is expressed by the formula $(2^n - 2n)x$, where:

- n = number of cycles;
- $2n$ = first product obtained after cycle 1 and second products obtained after cycle 2 with undefined length;
- x = number of copies of the original template.

Potentially, after 20 cycles of PCR there will be 2^{20}-fold amplification, assuming 100% efficiency during each cycle. The efficiency of a PCR will vary from template to template and according to the degree of optimization that has been carried out. *Table 1.1* shows a quantitative analysis of a PCR before and after 25 cycles and assumes a 10^5-fold amplification, this correlates to approximately a 70% efficiency of amplification. The target sequence product (also known as an amplicon) which is obtained contains the oligonucleotide primer sequences at its ends. Although extremely efficient, amplification of target sequences in an exponential manner is not an unlimited process. A number of factors act against the process being 100% efficient at each cycle. Their effect is more pronounced in the later cycles of PCR. Normally, the amount of enzyme becomes limiting after 25–30 cycles of PCR, which corresponds to about 10^6-fold amplification, due to molar target excess. The enzyme activity also becomes limiting due to thermal denaturation of the enzyme during the process. Another factor that will reduce the efficiency is the reannealing of target strands as their concentration increases. The reannealing of target strands then competes with primer annealing. It is important to minimize any variance in the PCR procedure and strive to obtain cycle efficiencies of 100%.

The basic components of a typical PCR are shown in *Figure 1.5* and are discussed in more detail later. The various reagents and equipment for PCR are readily available from many commercial suppliers and an extensive (but not exhaustive) selection is presented in Appendix B. Template DNAs may be provided by the researcher or clinician, alternatively genomic DNAs, genomic libraries, cDNA libraries and RNAs are also available commercially. Oligonucleotide primers can be routinely synthesized in many laboratories or can be purchased as custom synthesized reagents. PCR reaction components

TABLE 1.1: Quantitative analysis of a PCR before and after 25 cycles (amplification efficiency = ~70%)

	Before PCR				After PCR			
	Weight	Moles	Molarity	Molecules	Weight	Moles	Molarity	Molecules
Template[a]	1 ng	3.10×10^{-17}	3.10×10^{-13}	1.86×10^7	1 ng	3.00×10^{-17}	3.00×10^{-13}	1.81×10^7
Target[b]	10 pg	3.00×10^{-17}	3.00×10^{-13}	1.81×10^7	1 µg	3.00×10^{-12}	3.00×10^{-8}	1.81×10^{12}
Primers[c]	1623 ng	2.00×10^{-10}	2.00×10^{-6}	1.20×10^{14}	1574 ng	1.94×10^{-10}	1.94×10^{-6}	1.17×10^{14}
dNTPs[d]	39 µg	8.00×10^{-8}	8.00×10^{-4}	4.82×10^{16}	37 µg	7.70×10^{-8}	7.70×10^{-4}	4.64×10^{16}
Magnesium ion[e]	3.6 µg	1.50×10^{-7}	1.50×10^{-3}	9.03×10^{16}	3.6 µg	1.50×10^{-7}	1.50×10^{-3}	9.03×10^{16}
Taq DNA polymerase[f]	12.5 µg	1.33×10^{-13}	1.33×10^{-9}	8.01×10^{10}	12.5 µg	1.33×10^{-13}	1.33×10^{-9}	8.01×10^{10}

[a] Bacteriophage lambda (template dsDNA = 48 500 bp).
[b] Target is 500 bp.
[c] 1 µM (each) primers, 25-mers.
[d] 200 µM (each) dNTPs; total [dNTPs] = 0.8 mM (average molecular weight of a dNTP is 487 Da; average molecular weight of a dNMP is 325 Da).
[e] Total [$MgCl_2$] = 1.5 mM; free [$MgCl_2$] = 0.7 mM.
[f] 2.5 Units Taq DNA polymerase per 100 µl; polymerase activity = 250 000 units mg^{-1}; enzyme half-life not considered.

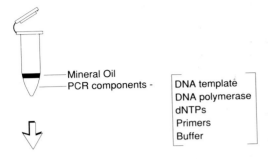

Mineral Oil
PCR components -
DNA template
DNA polymerase
dNTPs
Primers
Buffer

Thermal cycler
for amplification

FIGURE 1.5: *Basic components of a PCR amplification. Template DNA, buffer, dNTPs and primers are mixed and heated to denature the DNA thermally. DNA polymerase is added and mixed. The reaction components are overlaid with mineral oil and the tube placed in a thermal cycler. At the end of the amplification, the products are analyzed to confirm that the PCR has been successful.*

such as buffers, $MgCl_2$ and dNTPs are also available from many suppliers or can be prepared by the researcher. A wide range of thermostable DNA polymerases is now available from many manufacturers and the applications and properties of these are discussed in more detail in Section 2.2.

PCR was invented in 1985 by Kary Mullis [1], then working for Cetus Corporation in California, while driving late at night pondering new ways to detect specific bases by DNA sequencing. Indeed, Kary Mullis gives us a fascinating insight into what led to this invention in a *Scientific American* paper [2]. The 1993 Nobel prize for chemistry was awarded to Dr Mullis for having invented PCR. The original protocols for PCR amplification [1,3,4] used the Klenow fragment of *E. coli* DNA polymerase I to catalyze the oligonucleotide extension. However, this enzyme is thermally inactivated during the denaturation step of a PCR cycle and so the researchers had to add a fresh aliquot of enzyme at each cycle of the amplification process. Obviously this was a serious limitation, especially to attempts to automate the technique. The use of the Klenow enzyme worked well for the amplification of short fragments of DNA (<200 bp), but was disappointing for larger amplicons. Often the yields were low and the products showed size heterogeneity [5]. This was probably due to the low annealing and extension temperatures (37°C) which had to be used for the Klenow enzyme to be catalytically active. Another serious limitation in the early days of the PCR technique was the use of manual transfer of samples between water baths or heating blocks maintained at the specific temperatures required for each PCR cycle. Two crucial technological advances have made PCR such a key procedure in research and clinical laboratories:

(i) the use of thermostable DNA polymerases which greatly simplify the whole procedure because the addition of fresh enzyme after each denaturation step is no longer required. The first thermostable DNA polymerase to be introduced into PCRs was *Taq* DNA polymerase (see Section 2.2.1) [6];

(ii) the development of a variety of simple temperature cycling devices (also known as thermal cyclers or PCR machines) which led to the automation of the PCR (see Section 2.1).

The Polymerase Chain Reaction (PCR) process is covered by patents owned by Hoffmann-La Roche Inc. and F. Hoffman-La Roche, Ltd.

References

1. Saiki, R.K., Scharf, S., Faloona, F., Mullis, K.B., Horn, G.T., Erlich, H.A. and Arnheim, N. (1985) *Science,* **230,** 1350.
2. Mullis, K.B. (1990) *Scientific American,* April 1990, 56.
3. Mullis, K.B., Faloona, F., Scharf, S., Saiki, R., Horn, G. and Erlich, H.A. (1986) *Cold Spring Harbor Symp. Quant. Biol.,* **51,** 263.
4. Mullis, K.B. and Faloona, F. (1987) *Methods Enzymol.,* **155,** 335.
5. Scharf, S.J., Horn, G.T. and Erlich, H.A. (1986) *Science,* **233,** 1076.
6. Saiki, R.K., Gelfand, D.H., Stoffel, S., Scharf, S.J., Higuchi, R., Horn, G.T., Mullis, K.B. and Erlich, H.A. (1988) *Science,* **239,** 487.

2 Instrumentation, Reagents and Consumables

2.1 Instruments

In the evolution of the PCR two developments have greatly contributed to the success of the procedure: the use of a thermostable DNA polymerase (see Section 2.2) and instrumentation to allow the automation of temperature cycling. Two types of approaches to automating PCR have been devised, based on robotics or temperature cycling devices. In robotics methods the samples are mechanically moved between controlled temperature positions. There are disadvantages with this approach particularly in the large amount of bench space such instruments occupy, slow temperature equilibration if fine tuning is necessary and the difficulty of controlling temperature change when samples are being transported. Therefore temperature cycling devices have been the choice for PCR.

Several designs of temperature cycling instruments have been used, for example:

(i)   heating and cooling by fluids;
(ii) heating by electric resistance and cooling by fluid (e.g. tap-water or recycling water in a refrigerated bath) or refrigerant circulation;
(iii) heating by electric resistance and cooling by semiconductors (Peltier devices);
(iv) heating matt black metal surfaces by light followed by air cooling.

Thermal cycling devices are now available from many manufacturers (see Appendix B) and these have easy-to-use software for fully automated operation, together with accurate and reproducible temperature control in all sample positions in the cycler. A typical temperature cycling profile for a three-step protocol is shown in *Figure 2.1*. The thermal cycling parameters are critical to a successful PCR. The following parameters are listed as a guide to the important steps in the temperature and time profile of thermal cycling:

- denaturation,
- annealing of primers,
- extension of the primers,
- cycle number,
- ramp times.

Initial denaturation of template DNA at 95–100°C is sufficient to completely denature complex genomic DNA so that the primers can anneal after cooling. Thermal damage of DNA leads to an increased nucleotide misincorporation rate during PCR and high temperatures should be avoided if high fidelity is required [1]. However, supercoiled

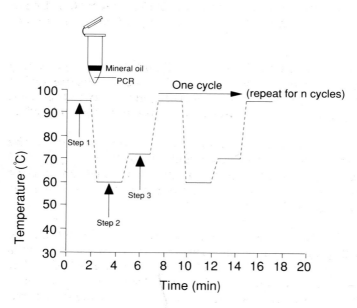

FIGURE 2.1: *PCR temperature cycling profile.* **Step 1:** *Heat denature the double-stranded DNA in the presence of primers, the four dNTPs, PCR buffer and a thermostable DNA polymerase. Denaturation is normally in the range 93–100°C.* **Step 2:** *Anneal the oligonucleotide primers to the denatured template by lowering the temperature to 37–65°C depending on the T_m of the oligonucleotide primers.* **Step 3:** *Extend the primers at 72°C with a thermostable DNA polymerase. Steps 1 to 3 constitute one cycle of PCR. This process is then normally repeated (cycled) for at least 20 cycles. At the last cycle the extension time may be increased by several minutes to ensure the complete synthesis of all strands.*

plasmid DNA substrates require boiling for several minutes to achieve complete denaturation. Denaturation during the PCR (i.e. cycle 2 onwards) is usually sufficient at 92–95°C but should be determined empirically as different PCR templates, thermal cyclers and tubes may have different requirements.

The primer annealing step is an important parameter in optimizing the specificity of a PCR, and calculation of T_m/annealing temperature is described in Section 2.4.2. The calculated annealing temperature is used as a starting point for experimental work; however, the annealing temperature should also be optimized empirically.

Primer extension is usually performed at 72°C, which is the optimum temperature for *Taq*/Amplitaq® DNA polymerase. The duration of the extension steps can be increased if long amplicons are being amplified, but for the majority of PCRs times of up to 2 min are usually sufficient. Often the extension time of the final cycle is longer (up to 10 min) in an attempt to ensure that all product molecules are fully extended.

The number of cycles is usually between 25 and 35. With increasing cycle numbers it is common to observe an increase in the amount of unwanted artifactual products and no increase in the desired product. Therefore it is unusual to find reactions that use more than 40 cycles.

The ramp time is that taken to change from one temperature to the next in the thermal cycler, and is dependent on the type of equipment used. Generally, the fastest ramps which can be attained are used for PCR. The actual ramp times can be determined by measuring the sample temperature with a thermocouple probe.

It is also possible to use a two-step protocol for PCR where the annealing and primer extension steps are both performed at a temperature up to 72°C and the denaturation step at 92–100°C. The temperature at which annealing and extension has to be performed is a compromise between the primers being able to anneal and the polymerase being sufficiently active to perform the extension.

Recent innovations in thermal cyclers include:

(i)   the use of 96-well microtiter formats, in place of microcentrifuge tubes, which allows robotic liquid handling devices to be used in sample preparation; 384-well plate thermocyclers are also available from Biometra;

(ii) the availability of temperature cyclers which allow different thermal cycles to be performed in independent blocks on the same machine;

(iii) temperature cyclers for the handling of microscope slides for use in *in situ* PCR experiments (see Section 12.5).

It is likely that future effort will be put into software development which will allow more complex programing and more flexibility in formatting of the samples, particularly when coupled to 96-well plate design, which is already extensively used in DNA sequencing protocols and for the storage of cosmid and yeast artificial chromosome (YAC) libraries. Another development is in the use of glass or plastic capillaries for PCR, either in a metal-block-type thermal cycler or in an air-cycling system. Reaction times for 30 cycles of PCR have been reduced to less than 30 minutes [2]. In summary, there are many commercially available dedicated thermal cyclers which allow accurate, reproducible and programable temperature control, and these have made PCR a routine procedure.

2.2 Enzyme choice

The introduction of thermostable DNA polymerases for PCR was one of the major technological advances which made PCR such a routine technique in research and clinical laboratories. Before the introduction of these thermostable enzymes, DNA polymerases such as the Klenow fragment of *E. coli* DNA polymerase I and T4 DNA polymerase, which are heat labile, were the only enzymes which could be used. This meant that fresh aliquots of enzyme had to be added at each PCR cycle. In contrast, the thermostable enzymes can be added in a single addition at the beginning of the amplification process without further additions during the reaction. *Table 2.1* lists some thermostable DNA polymerases which are currently in use for PCR, and the properties of these enzymes are shown in *Table 2.2*.

DNA polymerases are enzymes which catalyze the synthesis of long polynucleotide chains from monomer deoxynucleoside triphosphates using one of the original parental strands as a template for the synthesis of a new complementary strand. DNA synthesis always proceeds in the 5' to 3' direction since the polymerization is always from the 5' α-phosphate of the deoxynucleoside triphosphate to the 3' terminal hydroxyl group of the growing DNA strand. DNA polymerase, unlike RNA polymerase, requires a short DNA segment, or primer, to anneal to a complementary sequence and prime synthesis. Deoxynucleoside triphosphates (dNTPs) used in natural DNA synthesis normally comprise deoxyadenosine triphosphate (dATP), deoxycytidine triphosphate (dCTP), deoxyguanosine triphosphate (dGTP) and deoxythymidine triphosphate (dTTP). These

TABLE 2.1: *Thermostable DNA polymerases and their sources*

DNA polymerase	Natural/Recombinant	Source
Taq	Natural	Thermus aquaticus
Amplitaq®	Recombinant	T. aquaticus
Amplitaq® (Stoffel fragment)	Recombinant	T. aquaticus
Hot Tub™	Natural	Thermus flavis
Pyrostase™	Natural	T. flavis
Pfu	Natural	Pyrococcus furiosus
Pwo	Natural	Pyrococcus woesei
Tbr	Natural	Thermus brockianus
Tfl	Natural	T. flavis
Tli	Recombinant	Thermococcus litoralis
Vent™	Recombinant	T. litoralis
DeepVent™	Recombinant	Pyrococcus GB-D
Tth	Recombinant	Thermus thermophilus
UlTma™	Recombinant	Thermotoga maritima

TABLE 2.2: *Properties of the DNA polymerases commonly used in PCR*

	Taq/ Amplitaq®	Stoffel fragment	DeepVent™	Vent™	Pfu	Tth	UlTma™
Thermostability; half-life at 95°C (min)	40	80	400	1380	>120	20	>50[b]
5' → 3' exonuclease activity	Yes	No	No	No	No	Yes	No
3' → 5' exonuclease activity	No	No	Yes	Yes	Yes	No	Yes
Processivity	50–60	5–10	7	n.i.	n.i.	30–40	n.i.
Extension rate (nt sec⁻¹)	75	>50	>80	n.i.	60	>33	n.i.
Reverse transcriptase activity	Weak	Weak	n.i.	n.i.	n.i.	Yes	n.i.
Resulting DNA ends	3' A	3' A	>95% blunt	>95% blunt	n.i.	3' A	Blunt
Strand displacement	n.i.	n.i.	Yes[a]	Yes[a]	n.i.	n.i.	n.i.
Molecular weight (kDa)	94	61	n.i.	n.i.	92	94	70

Both Vent™ and *Pfu* are also available in an exonuclease minus (exo⁻) form.
[a] Strand displacement is temperature dependent.
[b] Measured at 97.5°C.
n.i. No information.

dNTPs attach to the free 3′-hydroxyl group of the primer and form a strand complementary to the template strand.

2.2.1 *Taq*/Amplitaq® DNA polymerase

The thermostable DNA polymerase from *Thermus aquaticus* (*Taq*) [3] has been the most extensively used enzyme in PCR. Other thermostable DNA polymerases have been isolated from other organisms and are commercially available for performing PCR (see *Table 2.1*). *T. aquaticus* was first isolated from a hot spring in Yellowstone National Park. *Taq* DNA polymerase is now available from many suppliers although Cetus Corporation was awarded a patent for both natural and recombinant *Taq* DNA polymerase as well as for the PCR process itself. *Taq* DNA polymerase has an optimal extension rate (polymerization rate) of 35–100 nucleotides per second at 70–80°C which is the optimum temperature range for the enzyme. Processivity, which is the average number of nucleotides incorporated before the enzyme dissociates from the DNA template, is relatively high for *Taq* DNA polymerase (see *Table 2.2*).

Taq DNA polymerase, and Amplitaq®, have a 5′ to 3′ exonuclease activity which removes nucleotides ahead of the growing chain. *Taq* DNA polymerase has been cloned and a modified version expressed in *E. coli* (Amplitaq®, P.E. Applied Biosystems) [4]. Amplitaq® has identical properties to *Taq* DNA polymerase (see *Table 2.2*) and is the most thoroughly characterized enzyme available for PCR. Since Amplitaq® is recombinant, the purity and reproducibility of this enzyme are higher than for *Taq* DNA polymerase, and Amplitaq® is the preferred enzyme for PCR. However, for PCR of bacterial targets containing DNA sequences homologous to those found in *E. coli* (which is the host used for the expression and production of Amplitaq®), it may be preferable to use *Taq* DNA polymerase rather than Amplitaq® to avoid potential contamination of the PCR with DNA from the enzyme itself. A quality-controlled version of Amplitaq® is available from P.E. Applied Biosystems Division especially for bacterial work, Amplitaq®, LD (low DNA). A modified version of the recombinant Amplitaq® DNA polymerase (Stoffel fragment, P.E. Applied Biosystems) is also available which has a deletion of 289 amino acids from the N terminus. This version has no intrinsic 5′ to 3′ exonuclease activity compared to the *Taq* or Amplitaq® DNA polymerases. The Stoffel fragment also has a twofold higher thermostability and exhibits optimal activity over a broader range of magnesium concentrations. Stoffel fragment is used particularly for improved amplification of templates known to be G–C rich, or to contain complex secondary structure, as the enhanced thermostability allows the denaturation temperature to be raised. The absence of 5′ to 3′ exonuclease activity allows more efficient amplification of circular templates such as plasmid DNA. Stoffel fragment exhibits optimal activity over a broad range of magnesium concentrations, which reduces the time spent on optimization of PCR conditions (see Section 2.3). It is particularly useful when performing multiplex PCR, where two or more targets are simultaneously amplified in the same reaction.

Taq, Amplitaq® and Amplitaq® Stoffel fragment DNA polymerases leave single 3′-dA nucleotide overhangs on their reaction products. If the PCR products have to be cloned, then these single base extensions have to be taken into account (see Chapter 4).

Amplitaq® GOLD (P.E. Applied Biosystems) is a recombinant thermostable 94 kDa DNA polymerase encoded by a modified form of the *T. aquaticus* gene. This form of the enzyme can be completely or partially activated in a pre-PCR heat step (hot start) or can be allowed to activate slowly during PCR (hot-start and time-release PCR). This prevents misprimed products and primer oligomers and should also result in higher yields of specific products.

2.2.2 Vent™ DNA polymerases

Vent™ DNA polymerase was first isolated from *Thermococcus litoralis*, which is a thermophilic archaebacterium found on ocean floors at temperatures of up to 98°C. The native form of *Tli* DNA polymerase is available (Promega) and the gene encoding this enzyme has also been cloned and expressed in *E. coli* (New England Biolabs). An exonuclease deficient derivative, Vent (exo⁻)™, and another version with higher thermostability (DeepVent™) are also commercially available (New England Biolabs). Vent™ DNA polymerase is more thermostable than *Taq* DNA polymerase and is capable of extending primers to give products up to 8–13 kbp in length. Vent™ polymerase possesses a 3′ to 5′ exonuclease activity that is responsible for the high level of fidelity, which is 5- to 15-fold greater than that of *Taq* DNA polymerase. DNA polymerases require the 3′ hydroxyl end of a base-paired polynucleotide strand (primer) on which to add further nucleotides. DNA molecules with a mismatched (i.e. not correctly base-paired) 3′-hydroxyl end can be corrected by the intrinsic 3′ to 5′ exonuclease activity of Vent™ DNA polymerase which will remove mismatched residues until a correctly base-paired terminus is generated. This is then an active template/primer substrate for the polymerase. Vent™ DNA polymerase also produces more than 95% DNA fragments without 3′ nucleotide additions (i.e.

blunt-ended fragments), which simplifies direct cloning of PCR products. Degradation of the primers due to the 3′ to 5′ exonuclease activity of this enzyme can be a problem. For this reason the enzyme should be added last when setting up the reaction. This problem can be overcome, however, by incorporating 3′-phosphorothioate linkages in the primers during their synthesis using standard chemical procedures [5].

2.2.3 *Pfu* DNA polymerase

This polymerase is isolated from the hyperthermophilic marine archaebacterium <u>P</u>yrococcus <u>fu</u>riosus (*Pfu*) and possesses both 5′ to 3′ DNA polymerase activity and 3′ to 5′ exonuclease proofreading activity [6]. The 3′ to 5′ exonuclease activity enhances the fidelity of DNA synthesis as it will excise incorrectly added, mismatched 3′-terminal nucleotides from the primer/template and then incorporate the correct nucleotide. The fidelity of DNA synthesis is 12-fold higher than that of *Taq* DNA polymerase. When setting up reactions with this enzyme it is essential to add the enzyme last (as for Vent™ DNA polymerase) since, in the absence of dNTPs, the 3′ to 5′ exonuclease activity of the enzyme results in degradation of template and primers.

Pfu DNA polymerase is available commercially from Stratagene, and a genetically engineered mutant of cloned *Pfu* DNA polymerase (exo⁻) is also available. The DNA polymerase specific activity is 10-fold higher than *Pfu* (exo⁺) but *Pfu* (exo⁻) has no detectable 3′ to 5′ exonuclease proofreading activity. The exo⁻ *Pfu* incorporates [α-^{35}S]dATP 10 times more efficiently than does *Taq* DNA polymerase and so has applications in [α-^{35}S]dATP cycle sequencing reactions (see Section 7.4), nucleotide analog incorporation and other reactions requiring a thermostable DNA polymerase lacking exonuclease activity.

2.2.4 *Tth* DNA polymerase

Five DNA polymerases have been isolated from <u>Thermus</u> <u>thermophilus</u> (*Tth*) and two forms have been shown to be efficient in PCR. Recombinant *Tth* DNA polymerase is a thermostable polymerase obtained by the expression in *E. coli* of a modified form of the DNA polymerase gene from *T. thermophilus* [7]. In the presence of manganese and at temperatures around 70°C, this polymerase can be used to reverse transcribe RNA efficiently. Subsequent PCR can be performed in the same tube using the intrinsic DNA polymerase activity simply by chelation of the manganese cation and the addition of magnesium (see Section 5.3). The use of *Tth* DNA polymerase for RNA PCR therefore has the advantage over conventional RNA (RT) PCR where a reverse transcriptase is used to synthesize cDNA and then a second reaction is required with a thermostable DNA polymerase for the PCR step.

The thermostability of the reverse transcriptase (RT) activity of *Tth* DNA polymerase is useful in amplifying DNA from RNA templates that contain G–C-rich sequences or secondary structures since the elevated temperatures serve to denature the template RNA. *Tth* DNA polymerase is also used in the detection and analysis of gene expression at the RNA level.

2.2.5 UlTma™ DNA polymerase

UlTma™ DNA polymerase is encoded by a recombinant modified form of the *Thermotoga maritima* DNA polymerase gene. *T. maritima* (*Tma*) is a hyperthermophilic, Gram-negative eubacterium that was first isolated from geothermally heated marine sediments off Italy. *Tma* grows naturally at temperatures up to 90°C. UlTma™ DNA polymerase exhibits greater thermostability than Amplitaq® DNA polymerase (see *Table 2.2*). This enzyme has no associated 5′ to 3′ exonuclease activity but does have an inherent 3′ to 5′ exonuclease proofreading activity and this can modify the 3′ end of primers when the enzyme, magnesium chloride and primers are combined for the PCR. This undesirable primer modification can be minimized by using hot-start techniques (see Sections 3.4 and 3.5). UlTma™ DNA polymerase should be used when high fidelity is required.

2.2.6 *Pwo* DNA polymerase

Pwo DNA polymerase is isolated from *Pyrococcus woesei* and possesses 3′ to 5′ exonuclease activity but no 5′ to 3′ exonuclease. The 3′ to 5′ exonuclease activity enhances the fidelity of DNA synthesis as it will excise any incorrectly added, mismatched 3′-terminal nucleotides from the primer extension product and then incorporate the correct nucleotide. The 3′ to 5′ exonuclease can also degrade primers or template and so it is advisable to add the enzyme last and use a hot-start method. This enzyme has a half-life of 2 h at 100°C and is useful for high temperature primer extension reactions, DNA cycle sequencing and blunt-end cloning. *Pwo* DNA polymerase is available commercially from Boehringer Mannheim.

2.3 Other PCR reagents

2.3.1 Deoxynucleoside triphosphates (dNTPs)

High-purity dNTPs are supplied by several manufacturers (see Appendix B) either as four individual stock solutions or as a mixture of all four dNTPs. Many stock solutions are now supplied already adjusted to pH 7.5 with NaOH. PCR is normally performed with dNTP concentrations around 100 µM, although at lower dNTP concentrations (10–100 µM) Taq DNA polymerase has a higher fidelity. However, the optimal concentration of dNTPs depends on:

- the $MgCl_2$ concentration,
- the reaction stringency,
- the primer concentration,
- the length of the amplified product,
- the number of cycles of PCR.

For optimization of a particular PCR it may be necessary to determine the best dNTP concentration empirically.

Some modified nucleotides have been incorporated into PCR products by the use of dNTP analogs. Some of the applications of these dNTP analogs are listed in *Table 2.3*.

2.3.2 Buffers and $MgCl_2$

There are several buffers available for PCR. The most common buffer used with Taq/Amplitaq® DNA polymerase has the following components in a 10× concentrated buffer and must be diluted 1 : 10 (v/v) prior to use:

- 100 mM Tris–HCl, pH 8.3 at room temperature,
- 500 mM KCl,
- 15 mM $MgCl_2$,
- 0.1% (w/v) gelatin.

Buffer compositions for use with other thermostable polymerases may differ; however, most suppliers usually provide a 10× buffer for use with the respective enzyme.

The $MgCl_2$ concentration in the final reaction mixture can be varied, usually within the range 0.5–5.0 mM in order to find the optimum. Mg^{2+} ions form a soluble complex with dNTPs which is essential for dNTP incorporation; they also stimulate the polymerase activity and

TABLE 2.3: *Nucleotide analogs that can be incorporated into PCR products, and their applications*

Nucleotide	Application
7 deaza-dGTP	Reduces DNA secondary structure where G-rich regions affect PCR efficiency and the resolution of sequencing reactions
dUTP	Substitution of dTTP by dUTP allows prevention of contamination after digestion with uracil *N*-glycosylase (UNG)
ddNTPs	Used in dideoxy DNA sequencing reactions
Fluorescent-dye-labeled ddNTPs	Used in automated fluorescent DNA sequencers
$[\alpha^{32}P]dNTPs$	Radiolabeling of PCR products
$[\alpha^{35}S]dATP$	Radiolabeling of PCR products
$[\alpha^{35}S]dCTP$	Radiolabeling of PCR products
Biotin-11-dUTP	Nonradioactive labeling of PCR products
Biotin-16-dUTP	Nonradioactive labeling of PCR products
Biotin-21-dUTP	Nonradioactive labeling of PCR products
Digoxigenin-11-dUTP	Nonradioactive labeling of PCR products
Bromodeoxyuridine	Nonradioactive labeling of PCR products
Inosine	Degenerate base for ambiguities in codons
Universal nucleotide	Degenerate base for ambiguities in codons

increase the T_m of the double-stranded DNA and primer/template interaction. The effect of varying $MgCl_2$ concentrations in a PCR is shown in *Figure 2.2*. The concentration of $MgCl_2$ can have a dramatic effect on the specificity and yield in a PCR. Concentrations of 1.0–1.5 mM $MgCl_2$ are usually optimal, but in some cases, different amounts of Mg^{2+} may prove to be necessary. Generally, insufficient Mg^{2+} leads to low yields and excess Mg^{2+} will result in the accumulation of nonspecific products.

2.3.3 Inhibitors and enhancers of PCR

PCR can be inhibited or enhanced by many different substances over and above those described previously. Many factors may lead to inhibition of PCR and we have only focused here on those that are likely to be experienced frequently when setting up PCRs. These may

t-PA

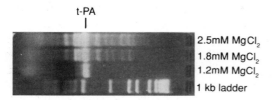

2.5mM MgCl$_2$
1.8mM MgCl$_2$
1.2mM MgCl$_2$
1 kb ladder

FIGURE 2.2: *The effect of MgCl$_2$ concentration on the specificity of PCR. Thirty cycles of PCR were performed with 50 ng of human genomic DNA and Amplitaq® DNA polymerase. The fragment being amplified is a 537 bp region from the tissue-type plasminogen activator gene (t-PA). Different concentrations of MgCl$_2$ were used in standard Amplitaq® PCR buffer. The size markers are 1 kb ladder (Gibco-BRL).*

arise from the nature of the native biological specimens and the method and reagents used to extract the DNA. Many different types of biological specimens are used for PCR, including animal tissues and bodily fluids, bacterial samples, forensic and archaeological material and plant tissues. Human DNA is typically obtained from peripheral blood cells, urine, fecal samples, cell smears, hair roots, semen, cerebrospinal fluid, biopsy material, amniotic fluid, placenta and chorionic villus. In addition to these sources, DNA from animals and birds may be extracted from tail sections and feather roots, respectively. Many of these crude preparations contain inhibitory substances which have not been identified, and PCR controls should be performed to ensure that inhibition is not occurring (see *Table 3.4*). One of the most common sources of material for extraction of DNA is blood, and it is best to prevent coagulation by collecting into EDTA tubes (1 mg ml^{-1}). Heparin, another commonly used anticoagulant, should not be used for collection of blood if PCR is to be performed on the extracted DNA, since heparin is a potent PCR inhibitor. Other substances in blood, probably porphyrin compounds, are also strong inhibitors of PCR and these can be eliminated from the DNA preparation by lysis of the red blood cells and centrifugation to pellet the white cells.

DNA extraction methods can be quite different for such a diverse group of biological specimens and this can cause difficulties in terms of the carry-over of different inhibitory agents from the extraction procedure into the PCR reaction. Extraction of DNA from different sources routinely uses detergents for cell lysis and denaturation.

Often, nonionic detergents (e.g. Triton X100, Tween 20, Nonidet P40) are used and these generally do not inhibit PCR at concentrations of up to 5%. On the other hand, ionic detergents such as sodium dodecyl sulfate (SDS) can only be tolerated at extremely low concentrations and therefore should be removed by phenol extraction and ethanol precipitation of the DNA before PCR. These detergents are normally used at concentrations higher than those that can be tolerated in PCR. For example, SDS is used at concentrations of 1–2% (w/v), but inhibits *Taq* DNA polymerase at concentrations higher than 0.01% (w/v). The inhibitory effects of low concentrations of SDS (e.g. 0.01%) can be reversed by certain nonionic detergents (e.g. 0.5% Tween 20 or Nonidet P40).

Many detergent extraction procedures are performed in the presence of proteinase K, a protease that will digest denatured proteins. *Taq* DNA polymerase is susceptible to protease digestion and so proteinase K must be removed or inactivated. Thermal denaturation at 95°C is sufficient to achieve this. Heat treatment is usually followed by phenol extraction which will also denature the proteinase K. Subsequent centrifugation will partition the protease into the phenol/organic phase, leaving the DNA in the aqueous phase. Residual traces of phenol, which also inhibits PCR, can then be removed by chloroform : iso-amyl alcohol (49 : 1) or ether extraction, or alternatively by ethanol precipitation of the DNA.

Many researchers have reported substances that can be added to a PCR to increase the efficiency or specificity. However, the mode of action has not been determined for any of these factors, although many possible explanations have been proposed. The major drawback of all of these substances is that no single one can be employed in any particular PCR with the guarantee of success. Many substances will enhance a PCR at a specific concentration but the transfer of these conditions to a different PCR may or may not result in an effect. Some enhancers of PCR are summarized in *Table 2.4*.

2.3.4 Template DNA

Template DNAs are usually provided by the researcher or clinician; however, a large number of genomic DNAs, genomic libraries (in lambda, cosmid or YAC vectors), cDNA libraries, total RNAs, poly(A)$^+$ RNAs (i.e. polyadenylated mRNAs, and cDNAs) are available commercially. Many of these are from different animal species and include a large variety of tissues and cell types. A number of plant DNAs, RNAs and libraries are also commercially available.

TABLE 2.4: *Enhancers of PCR*

Substance	Concentration
Formamide	5%
Dimethyl sulfoxide (DMSO)	<10%
Tetramethylammonium chloride (TMAC)	10–100 µM
Polyethylene glycol 6000 (PEG)	5–15%
Glycerol	10 –15 %
Tween® 20	0.1 –2.5%
Gene 32 protein (Pharmacia)	1 nM
7 deaza-dGTP	Replace 75% of dGTP with deaza-dGTP
Perfect Match® (Stratagene)	1 unit
Taq Extender™ PCR additive (Stratagene)	1 unit
E. coli single-strand DNA binding protein (ssb)	5 µg ml⁻¹

At higher concentrations these 'enhancers' are known to, or are likely to, inhibit *Taq* DNA polymerase, and so determination of the optimum concentration for enhancement has to be determined empirically.

The PCR sample may be single- or double-stranded DNA or RNA. If the starting material is RNA, either total RNA or poly(A)⁺ RNA can be used and first strand cDNA is prepared (see Section 5.3) prior to conventional PCR. Although the size of the DNA is generally not a prime factor, if high molecular weight genomic DNAs are being used then the amplification is improved if the DNAs are digested with a rare-cutting restriction enzyme (e.g. *Not*I or *Sfi*I). Often the concentration of the target sequence in the template DNA is not known and it is useful to optimize the PCR with positive control DNA. Normally subnanogram quantities of a cloned template and submicrogram quantities of genomic DNA are used for PCR and optimization trials. Potentially, PCR can be used to amplify as little as a single molecule of template, although great care has to be taken when performing such experiments, particularly in the avoidance of contamination (see Section 3.2).

The range of template DNAs (and RNAs) which have been used successfully for PCR is enormous and too large to detail here. There are references to many types of samples throughout this book, which will give the reader an idea of the scope of the technique.

2.4 Primers

2.4.1 Primer design

For most applications of PCR, the primers are designed to be exactly complementary to the template DNA. The design of the oligonucleotides is generally carried out using some simple guidelines, although several computer programs have been devised to aid primer design. In general, the primers used in PCR are between 20 and 30 nucleotides in length which allows a reasonably high annealing temperature to be used. There is no increase in specificity with primers longer than 30 nucleotides. The primers should, if possible, be made with an approximately equal number of each of the four bases, avoiding regions of unusual sequence such as stretches of polypurines, polypyrimidines or repetitive motifs. Sequences possessing significant secondary structure should also be avoided. Primer pairs should also be designed so that there is no complementarity of their 3′ ends either *inter* or *intra* individual primers. This precaution will reduce the incidence of 'primer-dimer' formation which is an amplification artifact caused when one primer is extended by the polymerase using the other primer or itself as a template, resulting in a short incorrect product.

For some applications the primers may not be exactly complementary to the template. For example, in site-directed PCR mutagenesis, the mutated base(s) is usually localized near the center of the oligonucleotide. Restriction endonuclease recognition sites (see Section 4.1.1), GC-clamps (see Section 10.1) or promoter sequences (see Section 4.2.1) are best incorporated at the 5′ end of the oligonucleotide, since the closer it is to the 3′ end of the primer the more likely it is that a mismatch will not be tolerated, so that the mismatched primer is not extended. In cases where degenerate primers have to be used, the degeneracy should be kept to a minimum at the 3′ end (see Section 5.5).

The distance between the primers when hybridized to the target DNA is usually less than 10 kbp. Indeed, a substantial reduction in synthesis efficiency is observed when amplicons exceed 3 kbp [8].

Where possible, primer design should also take into account features that will help to produce increased specificity (e.g. primers upstream or downstream of a coding sequence) or which allow the products to be detected (e.g. by internal restriction endonuclease sites or by hybridization with oligonucleotide probes). It may also be important to design amplification primers such that the PCR products can be definitively associated with DNA or RNA (e.g. PCR products which span an intron can be assigned to amplification of genomic DNA rather than RNA).

There are numerous commercial suppliers of the reagents for amplimer synthesis and also of custom amplimer synthesis. In addition, several suppliers have amplimer panels available (e.g. CLONTECH, P.E. Applied Biosystems and R&D Systems).

2.4.2 Melting temperature (T_m) of primers

The T_m can be calculated for a particular PCR primer using a number of equations, and for each amplification reaction it is best, if possible, to design pairs of primers such that each primer has a similar T_m. The most commonly used formula is [(number of A + T) × 2°C + (number of G + C) × 4°C], which was originally calculated in a 1 M salt concentration for oligonucleotide hybridization assays [9]. However, this is inaccurate with primers longer than 20 nucleotides (nt). Many laboratories use annealing temperatures of 3–5°C below the T_m calculated using this formula as a starting point for PCR optimization experiments.

Some researchers use the following equations to calculate primer T_ms:

(i) T_m = 81.5 + 16.6($\log_{10}$[J⁺]) + 0.41(%G+C) − (600/l) − 0.63(%FA), where [J⁺] = concentration of monovalent cations; l = oligonucleotide length; FA = formamide [10]. Normally PCRs are performed in the absence of formamide so the last term of the equation can be ignored. This formula is suitable for oligonucleotides of 14–70 residues.

(ii) T_p = 22 + 1.46(l_n), where T_p = optimized annealing temperature ± 2–5°C, l_n = effective length of primer = 2(number of G + C) + (number of A + T) [11]. This formula is suitable for oligonucleotides of 20–35 residues.

Formulae (i) and (ii) have been reported to be used to calculate the T_m of primers in the ranges of 14–70 nt and 20–35 nt, respectively. Suitable annealing temperatures are only approximately related to the T_m and the calculated T_ms only act as a reference point to begin experimentation. The ideal annealing temperatures may actually be 3–12°C higher than the calculated T_m. If routine PCR of a particular target is planned, then the optimum annealing temperature should be determined empirically. The highest annealing temperature which gives the best PCR products should be used.

There are also programs designed to help in primer design (e.g. OLIGO™, Primer Detective) which calculate the DNA duplex stability based on pairwise interacting nearest neighbor analysis. These programs also take into account other features such as regions of DNA complementarity and secondary structure which it is important to avoid.

2.4.3 Primer labeling

PCR products can be labeled via the primers used in the amplification reaction, either during or post amplification. Primers can be enzymatically labeled with ^{32}P by incubation with $[\gamma\text{-}^{32}P]ATP$ and T4 polynucleotide kinase or with ^{33}P using an equivalent reaction with $[\gamma\text{-}^{33}P]ATP$. The resulting PCR product can be detected easily and used for either direct sequencing by the Maxam and Gilbert method (see Section 7.7), DNA protein binding/protection assays, 'footprinting' or single-strand conformation polymorphism analysis (see Section 10.2).

PCR primers can be biotinylated at their 5′ ends, and the PCR products can be detected subsequently and quantitated non-isotopically, or can be captured by avidin or streptavidin, denatured and used for single-strand sequencing (see Section 7.5). There are a number of commercially available biotin phosphoramidites that allow primer biotinylation during the primer synthesis.

Fluorescent groups can also be attached to the primer(s), and by the use of different fluorescent tags on specific primer pairs it is possible to detect multiple targets in a single PCR reaction even if they co-migrate during electrophoresis. Fluorescein may be added directly to the primer during synthesis using a fluorescein phosphoramidite. Other fluorescent dyes, as active esters, are chemically coupled to the primer postsynthesis via a 5′-amino group incorporated by the addition of an amino phosphoramidite during primer synthesis. A 5′-amino group added in this way may also be used to label oligonucleotides with enzymes such as alkaline phosphatase. However, these enzyme tags do not retain their activity after PCR, due to heat denaturation. These enzyme-labeled oligonucleotides are therefore useful only as probes.

Developments in the chemistry of oligonucleotide synthesis have given rise to the ability to prepare non-nucleosidic phosphoramidite reagents which can be incorporated into PCR primers. Consequently these become incorporated into PCR products, and one result of this is the inability of *Taq* DNA polymerase to incorporate nucleoside triphosphates once the non-nucleosidic moiety is reached. These nonnucleoside moieties form impassable barriers to the polymerase and this gives rise to single-stranded tails on the PCR product [12]. The single-stranded tails can be any arbitrary or designed sequence and they essentially comprise a label on the PCR product. The hybridization properties of each tail are unique to the PCR product to which it was added. Thus each tail, by either hybridization to an immobilized oligonucleotide or a separately tagged oligonucleotide in solution, can be used for either capture or signaling purposes. For example, when an amplification refractory mutation system (ARMS, see Section 11.1) primer is made with a nonamplifiable tail which is exclusive to the allele it is to detect, the single-stranded tail can be used for capturing the product on a solid phase. A single-stranded tail on the common primer of the ARMS reaction that is both common to each allele and the control PCR product may then be used in a generic signaling system in conjunction with a single complementary detection oligonucleotide.

2.4.4 Calculating primer concentrations

To do this, it is necessary first to calculate the molar extinction coefficient of the primer at 260 nm. This can be calculated using the formula: $(8400 \times T) + (15\ 200 \times A) + (12\ 010 \times G) + (7050 \times C)$, where T, A, G and C are the number of times that each respective residue occurs in the primer. The molar extinction coefficient is equivalent to the absorbance at 260 nm (A_{260}) of a 1 M solution of primer. The absorbance of the primer solution is then measured at 260 nm (diluted 50-fold if the primer is crude rather than purified). Dividing the A_{260} of the primer stock solution by the molar extinction coefficient will give the molar concentration of the primer. Primers may be diluted to give a 20 µM working solution (it is often convenient when the same primer combinations are routinely used, to combine the primers in a mix where each primer is 20 µM within the mix, for subsequent dilution to up to 1 µM in the reaction mixture).

2.5 Consumables

Reaction vessels. These are normally 200–600 µl capacity and are now available in thin-walled versions which allow PCR cycles to be shortened due to the improved thermal transfer to and from the reaction. Several manufacturers use a 96-well format for the reaction tubes, there are also 96-well microtiter plates available specifically for PCR; 384-well plates and a thermal cycler with a 384-well block is available from Biometra.

Liquid handling consumables. Liquid handling consumables are also very important. Preferably, disposable pipette tips should be plugged to avoid aerosol contamination of the pipette barrel, alternatively, positive displacement pipette tips with disposable plungers should be used.

Mineral oil. A layer of light mineral oil overlaying the PCR is necessary to prevent evaporation in thermal cyclers which do not heat the lids of the reaction tubes. Normally, 50–60 µl of oil is added to a 100 µl PCR and the amount should be standardized to prevent variability in the efficiency of PCR, which can be caused by too much oil altering the thermal profile. The oil also helps prevent sample to sample contamination. Where the lid of the thermal cycler is heated (e.g. P.E. Applied Biosystems, thermal cycler GeneAmp 2400 or 9600; MJ Research, DNA engines; Techne, Cyclogene or Gene E) it is not necessary to overlay the reaction with oil. Silicone oil can be used as an alternative in place of mineral oil [13]. The oil can simply be removed by pipette after PCR, or can be removed by extraction with chloroform. Alternatively, a pipette tip can be inserted through the oil layer and an aliquot of the aqueous phase (the PCR) removed for analysis.

PCR gems. An alternative to an oil overlay to prevent evaporation and reduce contamination is the use of Ampliwax™ PCR gems (P.E. Applied Biosystems). These are precisely sized, specifically formulated wax beads. They can be used in place of mineral oil as a vapor barrier and may also help prevent contamination. However, they are particularly useful in the hot-start PCR method (see Section 3.4) where they can prevent the mixing of all of the PCR components until the reaction tube has reached a temperature at which nonspecific priming is minimized. This also synchronizes the start of the PCR in different thermal cycler positions.

References

1. Eckert, K.A. and Kunkel, T.A. (1991) in *PCR: A Practical Approach* (M.J. McPherson, P. Quirke and G.R. Taylor, eds). Oxford University Press, Oxford, p. 225.
2. Wittwer, C.T., Fillmore, G.C.H. and Garling, D.J. (1990) *Anal. Biochem.*, **186**, 328.
3. Saiki, R.K., Gelfand, D.H., Stoffel, S., Scharf, S., Higuchi, R., Horn, G.T., Mullis, K. and Erlich, H. (1988) *Science*, **239**, 487.
4. Lawyer, F.C., Stoffel, S., Saiki, R.K., Myambo, K., Drummond, R. and Gelfand, D.H. (1989) *J. Biol. Chem.*, **264**, 6427.
5. de Noronha, C.M.C. and Mullins, J.I. (1992) *PCR Methods Appl.*, **2**, 131.

6. Lundberg, K.S., Schoemaker, D.D., Adams, M.W., Short, J.M., Sorge, J.A. and Mathur, E.J. (1991) *Gene,* **108,** 1.

7. Myers, T.W. and Gelfand, D.H. (1991) *Biochemistry,* **30,** 7661.

8. Jeffreys, A.J., Wilson, V., Neumann, R. and Keyte, J. (1988) *Nucleic Acids Res.,* **16,** 10953.

9. Suggs, S.V., Wallace, R.B., Hirose, T., Kawashima, E.H. and Itakura, K. (1981) *Proc. Natl Acad. Sci. USA.,* **78,** 6613.

10. Sambrook, J., Fritsch, E.F. and Maniatis, T. (eds) (1989) in *Molecular Cloning: A Laboratory Manual (2nd edn).* Cold Spring Harbor Laboratory Press, Cold Spring Harbor, NY.

11. Wu, D.Y., Ugozzoli, L., Pal, B.K., Qian, J. and Wallace, R.B. (1991) *DNA and Cell Biol.,* **10,** 233.

12. Newton, C.R., Holland, D., Heptinstall, L.E., Hodgson, I., Edge, M.D., Markham, A.F. and McLean, M. (1993) *Nucleic Acids Res.,* **21,** 1155.

13. Ross, J.A. and Leavitt, S.A. (1991) *BioTechniques,* **11,** 618.

3 Amplifying the Correct Product

The overall efficiency and specificity of PCR is determined by many factors and several of these have been discussed in Chapter 2. The quality of the DNA template, the design of the primers (see Section 2.4.1) and the PCR conditions such as $MgCl_2$ concentration, DNA polymerase and buffer conditions are particularly important; these are listed along with the other factors in *Table 3.1*. PCR optimization kits are available commercially (e.g. from Boehringer Mannheim, Invitrogen and Stratagene); these can provide a good starting point for optimization. If a detailed PCR optimization strategy is necessary, then many variables have to be taken into account and this soon becomes a cumbersome analysis. Matrix analysis using Taguchi methods [1,2] can significantly simplify this procedure, and other experimental design techniques have also been reported [3].

3.1 Detection and analysis of PCR products

PCR products, or amplicons, consist of a fragment (or fragments) of DNA which is normally of a length defined by the boundaries of the

TABLE 3.1: *Optimization of PCR – critical parameters*

Parameters for optimization of PCR

PCR thermal profile (primer annealing, extension, denaturation, ramp times, number of cycles)
Magnesium ion concentration
Primer design and concentration
Template quality and concentration
PCR buffer
dNTP concentration
Addition of PCR enhancers
Avoidance of PCR inhibitors
Secondary PCR with nested primers
'Hot-start' PCR

PCR primers. Generally three types of information can be derived from the analysis of specific PCR products:

(i) detection of the presence of the target DNA sequence and variations within it;
(ii) quantitation of the yield of a PCR amplification to determine the relative or absolute amounts of the initial DNA or RNA target;
(iii) sequence analysis, either by differential probe hybridization or by direct sequencing of the product.

In some PCR methods, the length of the product is not known in advance and this will require some further characterization, for example in chromosome walking (see Chapter 8) or RACE-PCR (see Section 5.2.1). PCR products are often less than 10 kbp in length. Many techniques can be used to detect and confirm the identity of the amplified species (see *Table 3.2*). Normally the simplest and most commonly used method is electrophoresis.

3.1.1 Gel electrophoresis

Electrophoresis of an aliquot of the PCR on an agarose or polyacrylamide gel followed by visualization after staining with ethidium bromide, a fluorescent dye that intercalates into the DNA,

TABLE 3.2: Detection, identification and quantification of PCR products

Detection	Visualization
Agarose gel and/or polyacrylamide gel electrophoresis	EtBr staining (UV transilluminator, image analyzer)
	Southern blotting (hybridization with labeled probe)
	Incorporation of label into amplicon
	Addition of capture tag followed by detection
	Silver staining
Restriction endonuclease digestion	Agarose or polyacrylamide gel, HPLC
Dot blots	Hybridization with labeled probe (e.g. ASOs)
High-pressure liquid chromatography	UV detection
EtBr incorporation during PCR	UV transilluminator
Electrochemiluminescence	Voltage-initiated chemical reaction/photon detection
Scintillation proximity assay (SPA)	Scintillation counting of captured PCR product
Direct sequencing	Radioactive or fluorescent-based DNA sequencing

Specific methods for the detection of mutations are not shown, although modifications of those listed may be applied to mutation analysis.
Abbreviations: ASOs, allele-specific oligonucleotides; EtBr, ethidium bromide; HPLC, high-pressure liquid chromatography; UV, ultraviolet.

is one of the simplest methods of amplicon detection. After staining, ultraviolet transillumination allows visualization of the DNA in the gel. The gel may be photographed to provide a permanent record of the experimental result. Size markers and previously amplified control PCR products can be electrophoresed in adjacent wells of the gel to allow accurate size determination of the amplicons. An example of the wide size-range of specific amplification products is shown in *Figure 3.1*. It is also possible to detect several PCR products, as is required in multiplex PCR, using identical methods, provided the individual PCR products can be resolved. An example of the resolution of several PCR products from a multiplex PCR is shown in *Figure 3.2*. An alternative method of visualizing DNA bands in a gel is by silver staining, which shows the DNA directly without UV transillumination. A silver-stained gel of a multiplex PCR performed in the diagnosis of Duchenne muscular dystrophy is shown in *Figure 11.4*.

Standard DNA-grade agarose is usually sufficient for adequate resolution of PCR products. However, high-resolution agarose gels (e.g. NuSieve™, FMC Bioproducts) can be used to resolve DNA fragments of between 10 bp and 2 kbp, and are particularly useful in the resolution of small PCR products. For many small products, or in the detection of small size differences such as in microsatellite repeats

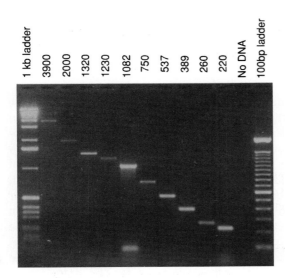

FIGURE 3.1: *PCR amplification of human genomic DNA with pairs of primers for different gene segments. The product sizes range from 220 bp to 3900 bp and the PCR product lengths are indicated at the top of each gel track. The size markers are 1 kb ladder and 100 bp ladder (Gibco-BRL).*

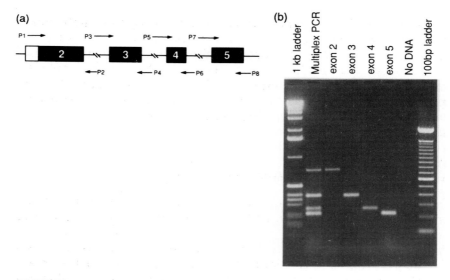

FIGURE 3.2: Multiplex PCR of four regions of the human alpha-1-antitrypsin gene. **(a)** Schematic representation of the structure of the human alpha-1-antitrypsin (AAT) gene with the coding exons 2–5 and the positions of the PCR primers (P1–P8). **(b)** Primer sets (P1–P8) flanking exons 2–5, of the human AAT gene were used for PCR. The four individual reactions are shown (indicated exon 2–exon 5 above each track) and the pool of all primer sets is also shown (multiplex PCR). Size markers are 1 kb ladder and 100 bp ladder (Gibco-BRL).

(see Section 9.3), nondenaturing or denaturing sequencing type polyacrylamide gels can be used.

Detection and/or confirmation of identity of a PCR product can be performed by Southern blot hybridization [4] or dot-blot hybridization (see Section 11.2) and detection with radioactive or nonradioactive specific probes. The Southern blot technique involves capillary blotting of the agarose or polyacrylamide gel on to a filter of nitrocellulose or nylon. The DNA is transferred by this process and, after immobilization by baking *in vacuo* (for nitrocellulose filters) or UV irradiation (for nylon filters), the filter can be hybridized with a specific probe. Southern blotting or dot blotting can enhance the sensitivity of detection compared with ethidium bromide staining, and can also confirm the identity of the product as being from the target gene. If oligonucleotide probes are used for the detection of PCR products, they are generally labeled with ^{32}P by incubation with [γ-^{32}P]ATP and T4 polynucleotide kinase. However, probes have been labeled with fluorescent dyes, digoxygenin, horseradish peroxidase, alkaline phosphatase, acridinium esters or other detectable tags.

Alternatively, PCR products can be labeled during the amplification reaction, for example by the addition of labeled dNTPs (see Section 2.3.1) or by addition of radiolabeled, fluorescent or biotinylated primers (see Section 2.4.3) to the reaction. Products can then be detected by gel electrophoresis and exposure to X-ray film or phosphorimager screen or by fluorescent detection using instruments such as the P.E. Applied Biosystems PRISM™ or 373A DNA sequencer.

3.1.2 Real-time amplicon detection

Real-time detection in the simplest form is achieved by the simultaneous amplification and detection of specific DNA sequences achieved by the addition of the DNA intercalating dye, ethidium bromide (EtBr), to a PCR [5]. Since the fluorescence of EtBr increases in the presence of double-stranded DNA, this can be used to detect double-stranded products after or during amplification. Amplification can be performed in the presence of EtBr with little or no effect on the yield or specificity of the PCR. Therefore it is possible to simultaneously amplify specific DNA sequences and detect the products of amplification by external monitoring on a UV trans-illuminator. This is a 'homogeneous' assay, since the reaction components do not have to be separated, and this simplifies PCR and may be applicable to high sample throughput.

Homogeneous assays can be performed using Perkin Elmer TaqMan™ Fluorogenic Probes in a 5′ nuclease assay. In this method a detection oligonucleotide probe is labeled at both the 5′ and 3′ ends with a fluorescent reporter and quencher molecule, respectively [6,7]. When the amplimers bind to their target sequence the detection probe is degraded by the 5′-exonuclease activity of *Taq* DNA polymerase. This releases the reporter from the vicinity of the quencher and a signal is generated which increases in direct proportion to the starting copy number. Dedicated instrumentation for this detection technique applying the 96-well microtiter format has been developed. Following the PCR the samples are transferred to a 96-well microplate and the amount of sequence-specific signal is measured by the TaqMan™ LS-50B luminescence spectrometer or the ABI PRISM™ 7200 Sequence Detection System. The ABI PRISM™ 7200 contains software for allelic discrimination, which allows automated detection of single base-pair differences with high throughput, and in addition, high-resolution mapping of genes can be performed using common single-base polymorphisms. The 7200 system can also streamline the detection of pathogens such as *Salmonella* and *Listeria* with rapid results obtained in hours rather than days.

The P.E. Applied Biosystems ABI PRISM™ 7700 Sequence Detection System, allows real-time quantitative PCR (see Chapter 13). In this system, a single instrument performs the thermal cycling and the detection. The latter uses a laser to induce fluorescence, and a charge coupled device (CCD) detector linked to real-time Sequence Detector software gives an analysis of the results. The increase in PCR products in all 96 wells can be measured as PCR proceeds and the results are available 1 min after completion of thermal cycling. The 7700 system also has a large linear dynamic range covering five orders of magnitude.

These systems also incorporate the PRISM™ fluorescent labeling reagents, allowing the detection of three accumulating amplicons simultaneously by employing three detection probes with 4,7,2′,4′,5′,7′-hexachloro-6-carboxyfluorescein (HEX), 4,7,2′,7′-tetrachloro-6-carboxyfluorescein (TET) or 6-carboxyfluorescein (FAM) as the 5′ fluorescein reporters, each with rhodamine as the 3′ quencher. By having the ability to use multiple labels in a multiplex PCR it becomes possible to run a positive internal control.

3.1.3 Immunological detection of PCR products

Enzyme-linked immunosorbent assay (ELISA) methods can be applied to the detection of PCR products. RNA probes can be hybridized to denatured amplicons in solution and subsequently followed by detection with a monoclonal antibody to DNA:RNA [8,9]. This is available commercially (SHARP Signal System, Digene Diagnostics Inc.). PCR is performed with one biotinylated and one nonbiotinylated amplimer, the product is denatured and hybridized in solution to a complementary unlabeled RNA. The DNA:RNA hybrids are captured via the biotin on the amplimer on to a streptavidin-coated microtiter plate and then detected using an alkaline phosphatase conjugated antibody specific for DNA:RNA hybrids after the addition of a colorimetric substrate. This system is reported to be approximately 100 times more sensitive than ethidium bromide staining of PCR products and can detect in the region of 10 pg of product. An analogous system (GEN-ETI-K™ DNA Enzyme Immunoassay, Sorin Biomedica) utilizes a biotinylated probe immobilized in the wells of a streptavidin-coated microtiter plate, denatured amplicons are hybridized to the probe and the hybridized dsDNA is then detected using an anti-dsDNA antibody, enzyme tracer and chromogenic substrate. This technology has been used to detect a range of genetic lesions and a variety of virus and parasite targets.

3.1.4 Scintillation proximity assay

The scintillation proximity assay (SPA) [10] is a unique technology commercially available from Amersham. SPA uses fluomicrospheres (SPA beads) coated with acceptor molecules which are capable of binding radiolabeled ligands in solution. A ligand is required which is labeled with an isotope that emits low-energy radiation (e.g. ^{3}H). This is dissipated easily into an aqueous medium. When labeled ligand is specifically captured by an SPA bead it will be close enough to activate the fluor and produce light, which can be detected by a scintillation counter. The majority of unbound labeled molecules are in an aqueous environment too far from the beads to be able to transfer energy to the fluor. Therefore SPA is a homogeneous assay and so does not require any further separation steps. The speed and convenience of SPA has been combined with PCR and allows the detection and quantification of amplicons. In the assay system, products generated by PCR can be quantified by two alternative methods. In the 'total amplimer' method, one of the PCR primers is biotinylated and [^{3}H] thymidine is incorporated into the amplicon during PCR. An aliquot of the PCR is added directly to streptavidin-coated SPA beads and analyzed by scintillation counting. In the 'specific amplimer' method, standard nonbiotinylated primers are used in conjunction with [^{3}H] thymidine to generate labeled amplicons. A 5′-biotinylated oligonucleotide which is complementary to a specific internal sequence in the target DNA is annealed to heat-denatured PCR products and then SPA beads are added and scintillation detected. This assay gives a good signal : noise ratio at low target concentrations and is linear over two orders of magnitude, making this a robust and flexible method for PCR quantification.

3.1.5 Electrochemiluminescent detection

Electrochemiluminescent detection and quantification of PCR-amplified DNA is also possible [11]. Tris(2,2′-bipyridine)ruthenium (II) chelate (TBR) in the phosphoramidite form (QPCR™ ruthenium phosphoramidite, P.E. Applied Biosystems) can be incorporated at the 5′ position of a PCR primer or hybridization probe during oligonucleotide synthesis. Within the electrochemical detection cell of the instrument (QPCR™ System 5000, P.E. Applied Biosystems), the reactions that lead to the emission of light from the ruthenium label are initiated electrically by applying a voltage to the sample solution. Several assay formats are possible. If there is only one specific product, then a TBR-labeled primer and a biotin-labeled primer are used for PCR. The product is captured by streptavidin-coated magnetic beads prior to electrochemiluminescent detection. Alternatively, one PCR primer can be labeled at its 5′ end with TBR. After PCR the products are denatured, hybridized with a biotin-labeled probe and captured on streptavidin beads. Capture of the PCR product by streptavidin-coated magnetic beads is rapid (*c*. 15 min) and hybridization, which takes place in solution, is also rapid (15 min). Electrochemiluminescence as a technique is specific with a low detection threshold (in the attomole range), which is equivalent in sensitivity to the use of radioisotopes. The TBR chelates are stable and able to withstand both temperature cycling, when used as tags on PCR primers, and the DNA heat denaturation step, if used as hybridization probes. This is an advantage over probes and primers which are labeled with enzymes or other heat labile reporter groups.

3.2 How to avoid contamination

There are many potential sources of pre-PCR contamination, some of which have only become apparent with the advent of such a sensitive technique as PCR. For example, if clinical samples have to be used, it is essential for thought to be given to the methods and materials used for collection and storage. Contaminating DNA may originate from any person's hair or skin cells that have been shed, and laboratory surfaces are potential sources of contamination. Air-borne contamination may be the cause of sporadic contamination of single samples. The reagents used for PCR or for other steps in the extraction/preparation of samples are also potential sources of extraneous DNA. For example, in the PCR reaction any of the buffer components, gelatin, oligonucleotide primers, dNTPs and even the DNA polymerase could be contaminated. It is also possible to encounter problems with other materials commonly used in sample extraction or preparation prior to PCR (e.g. reverse transcriptase, T4 DNA ligase, restriction enzymes, proteinase K). Indeed, unless reagents are guaranteed as PCR grade by the manufacturers then it is possible that they are contaminated with DNA. A list of possible contamination sources is provided in *Table 3.3*. Sporadic contamination cannot be controlled easily; however, repeated testing is unlikely to result in the same event occurring again [12].

It is important to perform control experiments with every PCR experiment, especially if PCR is being used as a diagnostic tool. *Table 3.4* lists a number of controls which should be considered when performing PCR. Negative controls should be designed to check the PCR reagents for contamination with DNA (e.g. genomic DNA, DNA

TABLE 3.3: *Potential sources of contamination in PCR*

Potential sources of contamination
Biological samples (e.g. from patients, animals, etc.)
Sample collection methods
Laboratory staff
Laboratory environment/air conditioning
Liquid nitrogen/ice
Tissue homogenizer
Pipettes/pipette tips
Reaction tubes/glassware
Reagents
Recombinant or biological products (e.g. gelatin, bovine serum albumin, restriction enzymes)
Hood or fume-cupboard filters
Centrifuges/centrifuge tubes
Microtome blades
Thermal cycler, heating blocks, water baths
UV transilluminator
Electrophoresis apparatus
Post-PCR contamination from the handling of PCR products (e.g. pipetting, gel loading, Southern blotting, etc.)

from infectious agents or previously obtained PCR products). Positive controls with well-characterized samples are also critical in determining the efficiency and specificity of PCR.

TABLE 3.4: *Useful controls for setting up PCR*

Control for	Method
Contamination of reagents with target DNA	PCR in the absence of exogenously added DNA
Presence of 'amplifiable' DNA in the sample	PCR with primers for alternative target
Reaction specificity	Use negative control to check for spurious background bands
Set-up, thermocycling, reaction sensitivity	Use negative and positive controls to check that PCR parameters are suitable and that expected product yields are obtained
Completeness of PCR mixture (checks whether essential components are missing or degraded)	PCR with positive control DNA
Result plausibility	Repeat PCRs with primers from an independent but related sequence
Inhibition from endogenous substances known in the sample	Spike control reaction with a known amplifiable target and its respective primers
Reliability of results and occurrence of sporadic contaminations	Perform PCR twice

To avoid contamination in PCR there are a number of steps, some of which are common sense whereas others involve specific protocols, which can be followed to minimize the problem. Ideally, experiments should be set up either in a laminar flow cabinet or in a separate laboratory or at least in a designated area of the laboratory.

Equipment and reagents for use in setting up PCRs should be kept as a separate supply from general laboratory equipment, and especially from reagents used for the analysis of PCR products. Any materials that can be autoclaved should be sterilized, preferably in an autoclave used exclusively for PCR equipment and reagents. Disposable gloves should be worn at all times, as well as standard safety spectacles and laboratory coats.

Reagents such as buffers and dNTPs should be prepared in batches and aliquoted for storage in convenient-sized lots to prevent contamination of stock solutions. As already mentioned above, unless a reagent is certified as PCR grade it is possible that it is contaminated with DNA. Indeed, even *Taq* DNA polymerase itself from several suppliers has been reported to contain prokaryotic and eukaryotic DNA [13]. The most potent sources of contamination are PCR products from previous reactions or positive-control plasmids. If positive controls are being used for reamplification, then these should be diluted in a separate dispensing area. With adherence to these good laboratory practices it is unusual to have persistent problems; however, there are specific methods which can help avoid PCR contamination. Modifications of liquid-handling devices which help prevent contamination are invaluable, such as positive-displacement pipettes and pipette tips containing barriers to prevent aerosols reaching the inside of the pipette itself.

3.2.1 Uracil *N*-glycosylase (uracil DNA glycosylase)

Uracil *N*-glycosylase (UNG), also known as uracil DNA glycosylase (UDG), can be used for decontamination of PCR reactions prior to initiating thermal cycling. *Taq* DNA polymerase has a similar affinity for dUTP as for dTTP. This allows the substitution of dUTP for dTTP in the PCR reaction. PCR products therefore have uracil incorporated and uracil is a substrate for UNG which catalyzes the hydrolysis of the *N*-glycosidic bond in nucleic acids containing uracil. Therefore when PCR reactions are pretreated with UNG any carry-over from previous dUTP-substituted PCR products is eliminated. However, some DNA polymerases do not incorporate dUTP as efficiently as *Taq* DNA polymerase [14], therefore with these enzymes UNG decontamination is not possible. UNG is available from Boehringer

Mannheim, Gibco-BRL, New England Biolabs and P.E. Applied Biosystems. A carry-over prevention kit using this method is also supplied by P.E. Applied Biosystems.

An alternative carry-over prevention method has been reported in which amplimers are synthesized with dUTP for dTTP substitutions, and so reamplification of PCR products from carry-over can be inhibited by UNG treatment [15].

3.2.2 UV irradiation

Contaminating DNA in PCR tubes can be damaged by UV irradiation and converted into a nonamplifiable form. Dry DNA requires longer exposures to UV light than DNA in solution [16]. Various other parameters influence the susceptibility to damage by UV such as:

- the pyrimidine content (particularly thymidine),
- the length of the target DNA,
- the irradiation time,
- the distance from the UV source,
- the wavelength,
- the energy.

UV irradiation for more than 10 min with 100 μJ min^{-1} at 254 nm in a UV crosslinker (e.g. Amersham, Stratalinker, Stratagene) should be used for reaction tubes and the reaction mixtures (without the sample DNA, primers and DNA polymerase) [17].

3.2.3 Enzymatic treatment

DNase I [18] or exonuclease III digestion [19] or treatment with a restriction enzyme which has recognition sites within the target DNA [18] can be used. However these have the drawbacks of adding extra steps to the PCR and may influence the efficiency of the reaction and may themselves pose contamination risks.

3.2.4 Psoralen and isopsoralen treatment

Psoralens (e.g. 8-methoxypsoralen (8-MOP)), intercalate into double-stranded nucleic acids. On irradiation with long-wave UV light (300–400 nm) they give rise to interstrand crosslinks. This property has been incorporated into a method to decontaminate pre-PCR reaction mixtures [20]. Pre-PCR incubation of the reagents with 8-MOP followed by UV irradiation does not affect the amplimers or *Taq* DNA polymerase but does render contaminating DNA unable to denature during PCR.

Isopsoralens, such as 4'-aminomethyl-4,5'-dimethylisopsoralen (4'-AMDMIP) and 6-aminomethyl-4,5'-dimethylisopsoralen (6-AMDMIP) are known to form covalent monoadducts with dsDNA when photochemically activated by long-wave UV light and this can be used as a method of post-PCR sterilization [21]. Reagents can be added prior to a PCR, and after an amplification the products are photoactivated, which results in damage to the DNA. DNA polymerases are blocked when they encounter a photochemically modified base in the DNA, so that if the damaged strand is carried over into a new reaction then it cannot act as a template for the PCR.

3.3 Reaction specificity

Conditions favoring enhanced reaction specificity:

- decreased amounts of Mg^{2+}, dNTPs, *Taq* DNA polymerase, primers,
- decreased numbers of cycles and shorter cycle segment lengths,
- increased annealing temperature,
- use of PCR enhancers,
- increased ramp speed,
- use of hot-start PCR,
- nested PCR,
- touchdown PCR.

The specificity of any PCR reaction is dependent on the specificity of primer hybridization to the targeted site with respect to nonspecific regions of the input DNA. Primers which undergo nonspecific extension in PCR not only reduce the concentration of primer available for the correct amplicon but can also lead to artifactual products, or false positive or negative results.

It is important to try to standardize the set-up procedures for PCRs and, if possible, reactions should not be left to stand around before being put into the thermal cycler. The hot-start procedure outlined below alleviates some of these problems. Many features of PCR affect the sensitivity and specificity of the reaction; these are covered in other sections of this book.

3.4 Hot start

The essential attribute of this method is the separation of one or more important components of the PCR such that all reaction components are mixed only after denaturation of the template [22]. Originally this was achieved by mixing the polymerase with a preheated reaction mixture. However, if large numbers of PCRs are to be set up, then it is difficult to have identical starting conditions for each tube. Furthermore, opening and closing tubes with an extra pipetting step presents a further risk of contamination. Therefore it is more efficient to separate the reaction components physically with a material that can be used as a barrier but which melts on raising the temperature, thereby causing mixing of all reaction components at the start of the PCR.

A precisely sized, specially formulated wax bead (Ampliwax™ PCR gems; see Section 2.5) is melted and rests on top of the incomplete reaction mixture (see *Figure 3.3*). When this is cooled the wax solidifies and the remaining components are added on top. The tubes are then placed in the thermal cycler and thermal cycling is commenced. Once the wax melts it rises above the aqueous components and causes mixing of all reagents. Thus all PCRs start simultaneously. PCR using the hot-start protocol is more specific and efficient than standard PCR [22], and hot start coupled with the use of PCR gems is more reproducible. The benefits of hot start are particularly evident with amplifications from low target copy

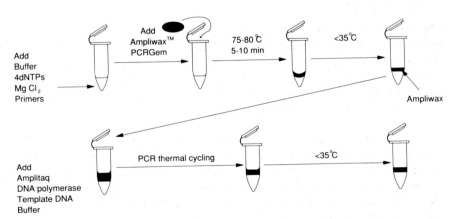

FIGURE 3.3: *The hot start technique with Ampliwax™. Components of PCR are added to the tube except DNA polymerase and template DNA. An Ampliwax™ PCR gem bead is added and melted at 75–80°C for 5–10 min then allowed to cool to solidify the wax. The remaining components, DNA polymerase, template DNA and buffer, are overlaid on the wax and the tube is placed in the thermocycler. Heating melts the wax, mixes the PCR components and the wax layer rises to the surface. At the completion of the PCR the tubes are cooled to <35°C and the wax solidifies.*

numbers. Because the reactants do not mix until the temperature is sufficiently high to melt the wax (55–58°C), this minimizes any nonspecific annealing of primers to nontarget DNA sequences and reduces the primer oligomerization incidence. Hot start may increase the performance in other applications such as multiplex PCR and the use of degenerate primers, where many primer sequences are present and each may potentially misprime. The solidified wax barrier has the advantage of protecting against spillage and evaporation, and, after removal of the sample by penetration of the wax barrier, it can be resealed by remelting and solidifying the wax.

3.5 Antibody-mediated hot-start PCR

A variation on hot start involves the addition of a neutralizing monoclonal antibody (TaqStart™ or TthStart™, CLONTECH) which specifically binds to *Taq* or N-terminal deletions of *Taq* DNA polymerase and *Tth* DNA polymerase, respectively. The antibodies block polymerase activity at ambient temperature but are themselves dissociated and denatured on heating during the first denaturation step of thermal cycling. They therefore help to reduce or prevent the generation of nonspecific amplification products and primer-dimer artifacts since the DNA polymerase activity is only restored at a temperature above which nonspecific hybridization of the primers is likely to occur.

3.6 Touchdown PCR

In touchdown PCR (TD-PCR) [23] cycling conditions are such that the desired amplicon is the preferred product and artifactual products and primer-dimers are minimized. The annealing temperature is lowered incrementally during the PCR cycling from an initial value above the expected T_m to a value below the T_m. This encourages the optimal hybridization of amplimer to target and the desired amplicon will begin to be amplified and accumulate before any undesired products. Most thermal cyclers can be programed to perform TD-PCR but the optimal cycling parameters have to be determined empirically.

3.7 Nested PCR

Nested PCR primers are ones that are internal to the first primer pair. The larger fragment produced by the first round of PCR is used as the template for the second PCR. Nested PCR can also be performed with one of the first primer pair and a single nested primer. The sensitivity and specificity of both DNA and RNA amplification can be dramatically increased by using the nested PCR method. The specificity is particularly enhanced because this technique almost always eliminates any spurious nonspecific amplification products. This is because after the first round of PCR any nonspecific products are unlikely to be sufficiently complementary to the nested primers to be able to serve as a template for further amplification, thus the desired target sequence is preferentially amplified. However, the increased risk of contamination is a drawback of this extreme sensitivity, and great care must be taken when performing such PCRs, particularly in a diagnostic laboratory.

3.8 Long-distance PCR (LD-PCR)

PCR over long distances (20–50 kb) is now possible [24–27]. LD-PCR uses a variety of systems and enzyme combinations which are commercially available from several suppliers. A prerequisite of the technique is the need for high quality genomic DNA template and this is best prepared using the protocols and procedures of genome researchers (see ref. 28 for example). Amplimer design should follow the same general principles as for standard PCRs but it may be important to match the melting temperatures of the two amplimers more accurately. A range of thermostable polymerases has been used succesfully for LD-PCR; Klentaq, a 5′-exonuclease deficient N-terminal deleted variant of *Taq* DNA polymerase in combination with *Pfu* DNA polymerase in a ratio of 180 : 1 provided the first LD-PCR [24]. Other polymerases (e.g. r*Tth*, *Taq* and *Tbr*) or combinations of polymerases (e.g. Klentaq and Vent™ or *Tth* and Vent™; Advantage™ (CLONTECH): r*Taq* and *Pfu* combination; TaqPlus Long™ PCR system (Stratagene): *Taq* and *Pwo*; Expand™ (Boehringer Mannheim): r*Tth* and Vent™; GeneAmp® XL (P.E. Applied Biosystems) have also been used for LD-PCR. These enzyme combinations together with other improvements in PCR conditions (e.g. touchdown PCR, hot start (see Sections 3.4–3.6)) and improved buffer systems have made PCR of targets up to 40 kb a routine procedure. The ability to amplify targets in the range 20–50 kb should revolutionize a broad range of PCR applications, particularly in genome research.

3.9 Key issues in setting up PCR

The key issues in setting up PCR are summarized in *Table 3.5* and these are discussed in more detail in Chapters 2 and 3. *Table 3.6* shows some of the common problems and challenges associated with PCR reactions and some possible solutions.

TABLE 3.5: *Key issues in setting up PCR*

Template
    DNA preparation (if LD-PCR then high quality genomic DNA is required)
    RNA preparation → cDNA synthesis for RT-PCR

Buffer composition
    Buffering agent, pH
    Mg^{2+}
    dNTPs
    KCl, $(NH_4)_2SO_4$
    PCR enhancers

Polymerase
    Choice for specificity, fidelity, yield and long-distance

Amplimers
    Design (length; GC content; T_m; inclusion of restriction sites, clamps, tags, etc.)
    Concentration in PCR

Thermal cycling parameters
    Denaturation, annealing and extension – temperature and length
    Two-step or three-step
    Touchdown PCR
    Hot-start PCR
    Nested PCR

Contamination avoidance
    Separate pre- and post-PCR
    Inclusion of appropriate positive and negative controls in all PCRs
    Specific contamination avoidance procedures

TABLE 3.6: *Troubleshooting guide*

(A) No PCR product observed

Reaction mixture incomplete or degraded	Always perform a positive control to check each reaction component
Insufficient cycles	Increase the number of cycles
Annealing temperature too high	Calculate the T_m of the amplimers and decrease the annealing temperature by a few degrees at a time
Not enough template	Use higher concentration of DNA (or RNA) and check quality of template
Suboptimal denaturation temperature and length	Optimize denaturation temperature and time by increasing or decreasing increments
Suboptimal extension time	Increase by 1 min increments, especially for long-distance PCR
Suboptimal reaction components	Mg^{2+}, dNTPs, amplimer concentrations require alteration
Difficult PCR	Some PCR reactions may require the addition of PCR enhancers (see Section 2.3.3)
Poor amplimer design	Check design of primers, e.g. sequence, GC content, length, T_m, lack of complementarity

(B) Multiple products

Too many cycles	Reducing the number of cycles may reduce nonspecific bands
Annealing temperature too low	Increase incrementally
Hot-start or touchdown PCR required	Use barrier-mediated/antibody-mediated / or temperature-mediated 'hot-start' or 'touchdown' PCR to increase specificity

(C) PCR products smeared

Too many cycles

Denaturation temperature too low

Extension time too long

Template degraded

Requirement for hot start or touchdown

Too much enzyme, Mg^{2+} or template

References

1. Taguchi, G. (1986) *Introduction to Quality Engineering.* Asian Productivity Organisation, UNIPUB, New York.
2. Cobb, B.D. and Clarkson, J.M. (1994) *Nucleic Acids Res.,* **22**, 3801.
3. Boleda, M.D., Briones, P., Farres, J., Tyfield, L. and Pi, R. (1996) *BioTechniques,* **21**, 134.
4. Southern, E.M. (1975) *J. Mol. Biol.*, **98**, 503.
5. Higuchi, R., Dollinger, G., Walsh, P.S. and Griffith, R. (1992) *Biotechnology,* **10**, 413.
6. Holland, P.M., Abramson, R.D., Watson, R. and Gelfand, D.H. (1991) *Proc. Natl Acad. Sci. USA,* **88**, 7276.
7. Livak, K.J., Flood, S.J.A., Marmaro, J., Giusti, W. and Deetz, K. (1991) *PCR Methods Appl.,* **1**, 142.
8. Bobo, L., Coutlee, F., Yolken, R.H., Quinn, T. and Viscidi, R.P. (1990) *J. Clin. Microbiol.,* **28**, 1968.
9. Lazar, J.G. (1993) *Am. Biotechnol. Lab.,* **11**, September, 14.
10. Bosworth, N. and Towers, P. (1989) *Nature,* **341**, 167.
11. Wages, J.M. Jr., Dolenga, L. and Fowler, A.K. (1993) *Amplifications,* no. 10, 1.
12. Kwok, S. and Higuchi, R. (1989) *Nature,* **339**, 237.
13. Schmidt, T.M., Pace, B. and Pace, N.R. (1991) *BioTechniques,* **11**, 176.
14. Slupphaug, G., Alseth, I., Eftedal, I., Volden, G. and Krokan, H.E. (1993) *Anal. Biochem.,* **211**, 164.
15. Longo, M.C., Berninger, M.S. and Hartley, J.L. (1990) *Gene,* **93**, 125.
16. Fairfax, M.R., Metcalf, M.A. and Cone, R.W. (1991) *PCR Methods Appl.,* **1**, 142.
17. Ou, C.-Y., Moore, J.L. and Schochetman, G. (1991) *BioTechniques,* **10**, 442.
18. Furrer, B., Candrian, U., Wieland, P. and Lœthy, J. (1990) *Nature,* **346**, 324.
19. Zhu, Y.S., Isaacs, S.T., Cimino, G.D. and Hearst, J.E. (1991) *Nucleic Acids Res.,* **19**, 2511.
20. Jinno, Y., Yoshiura, K. and Nikawa, N. (1990) *Nucleic Acids Res.,* **18**, 6739.
21. Cimino, G.D., Metchette, K.C., Tessman, J.W., Hearst, J.E. and Isaacs, S.T. (1991) *Nucleic Acids Res.,* **19**, 99.
22. Chou, Q., Russell, M., Birch, D.E., Raymond, J. and Bloch, W. (1992) *Nucleic Acids Res.,* **20**, 1717.
23. Don, R.H., Cox, P.T., Wainwright, B.J., Baker, K. and Mattick, J.S. (1991) *Nucleic Acids Res.,* **19**, 4008.
24. Barnes, W.M. (1994) *Proc. Natl Acad. Sci. USA,* **91**, 2216.
25. Foord, O.S. and Rose, E.A. (1994) *PCR Methods Appl.,* **3**, S149.
26. Cheng, S., Fockler, C., Barnes, W.M. and Higuchi, R. (1994) *Proc. Natl Acad. Sci. USA,* **91**, 5695.
27. Cheng, S., Chang, S.-Y., Gravitt, P. and Respess, R. (1994) *Nature,* **369**, 684.
28. Sambrook, J., Fritsch, E.F. and Maniatis, T. (eds) (1989) *Molecular Cloning: a Laboratory Manual (2nd edn).* Cold Spring Harbor Laboratory Press, Cold Spring Harbor, NY.

4 Cloning and Modification of PCR Products ✓

PCR has allowed the amplification of any region of even highly complex genomes in a few hours. Cloning of PCR products allows generation of relatively large amounts of the amplified region which avoids having to repeat the reaction every time the product is needed. Therefore when the product will be used as a probe or a positive control in further PCR experiments it is convenient to clone the amplicon. Similarly, a PCR product may be used for further studies (e.g. expression or mutagenesis) where it is necessary to clone the product. A number of cloning strategies are discussed in this chapter. It is important to plan a cloning strategy before the actual PCR is performed because many of these techniques require specific modifications designed into the amplimers. It is also recommended to analyze the amplicon to verify the size of the product and to determine its purity before its cloning is attempted.

Almost any DNA sequence can be engineered and clones screened using PCR as a replacement for the conventional techniques of recombinant DNA technology, with considerable savings in time, effort and expense. This is possible because of the ease and robustness of PCR and also the ability to choose the design of the amplimers used for PCR and to modify these with additional sequences if required.

4.1 Cloning of PCR products

4.1.1 The introduction of restriction sites

A restriction enzyme recognition site (or sites) can be added to the 5′ end of one or both oligonucleotide primers used for PCR [1] (see *Figure*

4.1). These sequences will be incorporated into the amplified product and can then be digested with the appropriate restriction enzyme to generate blunt or cohesive ends, as required. By the use of different restriction sites at either end of the amplicon, at least one of which results in a 'sticky' end, it is possible to clone the PCR product directionally into a vector.

Since the specificity of the primer is determined mainly by its 3' end, a short sequence towards the 5' end containing a restriction site has little effect on the specificity or efficiency of amplification. However, this method can sometimes be problematic:

(i) the restriction site(s) used in the PCR primer(s) may be present in the amplified fragment;

(ii) the restriction enzyme sites built into products via the primers are often difficult to digest unless extra nucleotides are added to the 5' end; and

(iii) sequential digestions must be performed in cases where restriction enzyme buffers are not compatible with each other.

This technique has been widely used and provides one of the most simple and reliable methods of cloning PCR products.

```
          5'
          C
          G
          C
           A
            G
            A
            A   EcoRI
           T
           T
               CCCTGCTCAGAAGCCACCATGGCTCT  3' → Direction of synthesis
3'GGGGTCGGGGGGCACGGGACGAGTCTTCGGTGGTACCGAGATACGGATACGTAGACATT →5'
                              |
                       PCR Amplification
                              ↓
           EcoRI
    5'CGCAGAATTCCCTGCTCAGAAGCCACCATGGCTCTATGCCTATGCATCTGTAA →3'
    3'GCGTCTTAAGGGACGAGTCTTCGGTGGTACCGAGATACGGATACGTAGACATT →5'
                              |
                      Digestion with EcoRI
                              ↓
        5'AATTCCCTGCTCAGAAGCCACCATGGCTCTATGCCTATGCATCTGTAA →3'
        3'GGGACGAGTCTTCGGTGGTACCGAGATACGGATACGTAGACATT →5'
```

FIGURE 4.1: *Cloning by the introduction of restriction enzyme recognition sites. The position of a PCR primer relative to the target DNA sequence is shown. The PCR primer will anneal to the target sequence and be extended in the 5' to 3' direction. An EcoRI restriction enzyme site has been added to the 5' end of the primer. Although this region does not match the template, the DNA can be amplified efficiently and the product will contain the restriction site at its end. Extra bases are normally added 5' to the restriction site to ensure that the restriction enzyme functions efficiently.*

4.1.2 Blunt-end cloning

Early attempts to clone PCR products simply involved their ligation into linearized, blunt-ended plasmid vectors. However, transformation frequencies were low. Various modifications were reported to improve the cloning frequency: 'polishing' the ends of the PCR products with an additional DNA polymerase (e.g. *Pfu* DNA polymerase, Klenow fragment of *E. coli* DNA polymerase I, T4 DNA polymerase) or exonuclease treatment (e.g. exonuclease III) and proteinase K treatment followed by phenol extraction to deproteinize the PCR products. Also, various methods to purify the PCR products before ligation have been reported that enhance the efficiency of blunt-end cloning. The transformation frequencies were still lower than expected for the amount of DNA being cloned. *Taq*/Amplitaq® DNA polymerase adds a single nontemplate specified nucleotide, almost always an adenosine (A) residue, to the 3′ ends of DNA [2] and so 'polishing' steps described above have to be taken to allow the PCR product to be cloned efficiently. Blunt-end cloning of PCR products generated using *Taq* DNA polymerase is becoming less common as improved cloning methods have become available. However, with the introduction of thermostable DNA polymerases (e.g. Vent™), which do not add additional bases to the 3′ ends of DNA, it is possible to blunt-end clone the PCR fragments without any pretreatment. There are vectors commercially available for blunt-end cloning (e.g. pCR Script™ and pCR Script™ Direct, Stratagene) and PCR products can be cloned either bidirectionally if one of the amplimers is 5′-phosphorylated, or unidirectionally with or without amplimer phosphorylation.

4.1.3 T–A vectors

The single A residues at the ends of a PCR product can serve as a one-base overhang to facilitate ligation, when the complementary thymidine (T) nucleotide is added to the 3′ ends of the cloning vector [3]. The orientation of the DNA insert cannot be controlled, and precautions have to be taken to avoid nonrecombinants due to vector molecules escaping T addition or false recombinants due to contaminating exonucleases removing the overhanging T residue. However, 'T–A' vectors that can avoid these drawbacks are available commercially, and the one-base sticky-end ligation is more efficient than blunt-end ligation. *Figure 4.2* shows the principle of 'T–A' vector cloning.

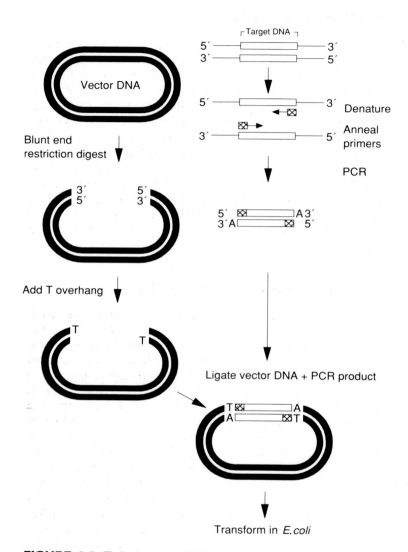

FIGURE 4.2: *T–A cloning of PCR products. Vector DNA is digested with a blunt-end generating restriction enzyme and a single T residue is added to the 3' ends using* Taq *DNA polymerase and dTTP. A normal* Taq *DNA polymerase PCR reaction produces a product with a single A residue added to each 3' end. The T–A overhangs in vector/PCR product, respectively, facilitate the ligation reaction, and the ligated vector-insert DNA is transformed into* E. coli.

4.1.4 Generation of half-sites

Generation of restriction half-sites has been used successfully, although this procedure has many steps [4]. Fragments are generated by a normal PCR reaction, using oligonucleotide primers containing the three 3' nucleotides of a six-base recognition site at their 5' ends. PCR products synthesized using *Taq*/Amplitaq® DNA polymerase require treatment with Klenow fragment of *E. coli* DNA polymerase I to remove the additional 3'-terminal A residue. Those generated by DNA polymerases which do not add a nontemplated residue (e.g. Vent™) can be used directly. The fragments are phosphorylated using T4 polynucleotide kinase and ATP, and are ligated into concatemers that are made up of two or more blunt-ended ligation products. The ligated products contain a six-base recognition sequence which can be cleaved with the appropriate restriction enzyme and cloned into a suitable vector.

4.1.5 Ligation-independent cloning (LIC)

PCR products can be cloned efficiently into a linearized, PCR amplified vector which has identical 5' end, single-stranded tails, which are not complementary to each other and which prevent the vector from recircularizing [5–7]. The sequence to be cloned is first amplified by PCR, using primers having complementary sequences to those that have been added in the vector. In the example shown (*Figure 4.3*) the PCR products are treated with T4 DNA polymerase in the presence of dGTP, which ensures that the exonuclease activity of this enzyme will not remove any bases beyond the first dG residue that is encountered, since the polymerase activity will replace the dG as rapidly as it is removed. The plasmid vector is PCR amplified with primers that incorporate a sequence complementary to those that have been added to the sequence to be cloned. In this case, the plasmid PCR product is treated with T4 DNA polymerase and dCTP. The PCR products are then annealed to the vector at room temperature for 30–60 min rather than being covalently joined by an enzymatic ligation step. An advantage of this method is that transformation efficiencies are almost 100%. It is possible to clone directionally using a similar procedure where different specific nucleotide sequences are incorporated into the 5' ends of the amplified fragments. Therefore the vector itself cannot recombine, but in the presence of the amplified gene of interest the fragment will anneal directionally and result in transformation. This cloning system is available commercially (PCR-Direct™ Cloning System, CLONTECH Laboratories Inc.).

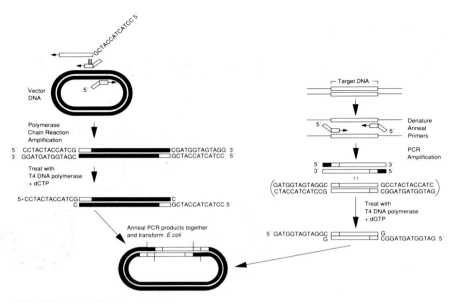

FIGURE 4.3: *Ligation-independent cloning. PCR products are generated from a plasmid vector and the target gene of interest using PCR primers designed to incorporate specific 13-nucleotide sequences into the 5′ ends of each of the amplified fragments. Amplified products are digested with the 3′ to 5′ exonuclease activity associated with T4 DNA polymerase in the presence of dCTP (vector PCR product) or dGTP (target gene PCR product). This creates 12-nucleotide single-stranded ends. Since the 5′ overhanging ends are complementary and sufficiently long to anneal stably at room temperature, the two treated products are mixed and used to transform* E. coli *without requiring joining with DNA ligase. The joining of the fragments is achieved* in vivo *after transformation.*

4.1.6 UNG cloning

With this method (see *Figure 4.4*), primers are designed that contain dUMP among the 5′-terminal nucleotides, these correspond to complementary overhangs of the vector sequences [8]. UNG cleaves the *N*-glycosidic bond between uracil and deoxyribose to yield abasic residues leading to disruption of base pairing. By generating a 3′-tail that contains at least 12 bases ensures efficient and stable annealing between product and vector DNAs. After PCR the amplicons are treated with UNG in the presence of the vector, annealed for 30 min at 37°C and then used to transform competent *E. coli* cells. Specific vectors have been designed for unidirectional and bidirectional cloning and are commercially available (CloneAmp® series, Life Technologies).

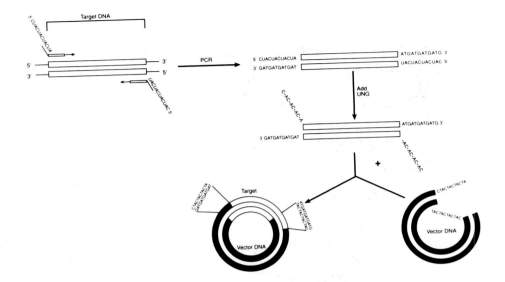

FIGURE 4.4: *UNG cloning strategy. dUMP PCR primer tails (12 nucleotides) are added to the 5' ends of synthesized amplimers as shown. Amplicons are digested with UNG which removes uracil, disrupting base-pairing and exposing 3' overhangs. These overhangs are annealed to complementary vector ends, and transformed into competent* E. coli. *Several vectors are available commercially (Life Technologies) and the vector illustrated represents CloneAmp® pAMP1. Both directional and nondirectional cloning vectors are available by having asymmetrical or symmetrical 12 nucleotide overhangs, respectively.*

4.1.7 pCR-Script™ cloning

Blunt-ended amplicons, for example those generated using *Pfu* polymerase or amplicons generated using *Taq* polymerase (or other nonproofreading or low proofreading polymerases) and then polished using *Pfu* can be cloned into pCR-Script™ vectors (Stratagene). This method allows rapid and efficient cloning of amplicons using bidirectional or directional methods. *Figure 4.5* shows the bidirectional method using a vector (e.g. pCR-Script™) with a rare cutting *Srf*I site that produces blunt-ended products. Blunt-end cloning is facilitated by using *Srf*I in the ligation mix to recut vector that self-ligates, and then transforming into *E. coli*.

Directional cloning can be performed by creating a monophosphorylated vector (pCR-Script™ Direct) by enzymatically treating the *Srf*I-digested vector with alkaline phosphatase to remove the 5' terminal phosphates, and subsequently digesting the plasmid

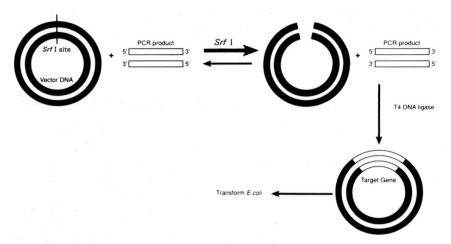

FIGURE 4.5: *pCR-Script™ cloning method. A PCR amplicon, together with a cloning vector which contains a unique site for the restriction enzyme SrfI, is added to a ligation reaction which contains T4 DNA ligase and also SrfI. This optimizes cloning by recutting vector which self-ligates. This mixture is then used to transform E. coli to obtain nondirectional clones. A modification of this method using pCR-Script™ Direct vector can be used for directional cloning (see Section 4.1.7).*

with a second restriction enzyme which generates blunt ends (e.g. *Sma*I). The PCR product to be cloned is generated using one 5′ phosphorylated amplimer (by kinase treatment or oligonucleotide synthesis) and one nonphosphorylated primer. After ligation a single-nicked circular molecule will be obtained in the desired orientation and a linear molecule will be obtained in the undesired orientation. RecBC strains of *E. coli* are transformed by linear DNA at very low efficiency and so directional cloning in the desired orientation is obtained.

4.2 Modification of PCR products

PCR is a reliable and rapid technique for DNA amplification and, because of the ease of addition of 'extra' sequence at the 5′ ends of primers, PCR offers an easy method for DNA sequence modification. The addition of these extra sequences has no effect on the efficiency or the specificity of amplification. The 3′ ends of the modified primers, which are complementary to the target, provide the necessary specificity, and the 5′ ends, although not matching the target, are copied during PCR and become part of the final PCR product. Indeed relatively large sequences, 45 nt or more, have been added successfully to the 5′ end of a primer and incorporated into the PCR product.

4.2.1 Adding promoters and ribosome-binding sites

Promoters are specific DNA sequences to which RNA polymerases bind and initiate transcription of the template into RNA. *In vitro* transcription is normally performed by insertion of the gene to be transcribed into a suitable vector downstream of an appropriate promoter, followed by transcription using the RNA polymerase specific for that promoter. Transcription vectors are available that contain SP6, T7, T3 or other bacteriophage promoters upstream from a multiple cloning region. Because additional DNA sequence can be added to PCR primers at their 5′ ends, it is possible to incorporate a promoter sequence (see *Figure 4.6a*) and to perform *in vitro* transcription directly on the PCR product [9,10]. Incorporation of an untranslated leader sequence into the PCR primer, between the promoter and the start of the gene of interest, can provide a suitable region for ribosome binding and initiation of protein synthesis. Thus *in vitro* transcription and translation can be performed using a PCR product simply by addition of the necessary signals to the appropriate primer. Messenger RNA that is transcribed can be studied directly or translated *in vitro* using, for example, rabbit reticulocyte or wheat-germ lysate, and a functional protein produced. This application of PCR has been named expression-PCR (E-PCR). Alternatively, sequences required for protein translation and subsequent cloning alone are incorporated into the primers so that PCR of the target DNA results in the synthesis of an expression cassette (expression-cassette PCR; EC-PCR)[11] (see *Figure 4.6b*). This expression cassette can then be cloned in an overexpression plasmid that provides transcriptional sequence elements such as a strong regulated promoter (e.g. *tac*) and an efficient transcription terminator (e.g. rrnBT$_1$T$_2$). The resulting construct is used to transform *E. coli* and high-level protein biosynthesis can be obtained.

The potential also exists to add other sequence motifs to the PCR primers, for example periplasmic signal sequences required for the export of proteins or Kozak sequences for efficient translation in eukaryotic expression systems, or peptide tags for purification and detection purposes may be incorporated. The length of the sequences that can be added to primers is primarily limited by the length of sequence that can be synthesized on automated DNA synthesizers and not by the PCR process itself. This problem can be circumvented

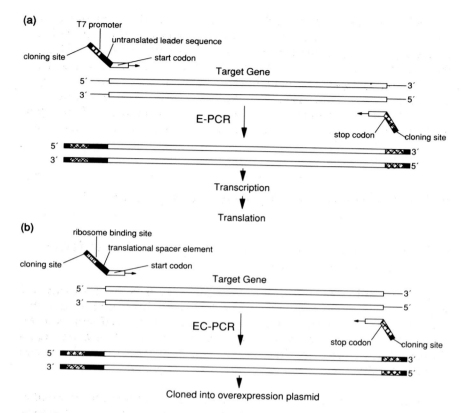

(a)

T7 promoter

untranslated leader sequence

cloning site

start codon

Target Gene

E-PCR

stop codon

cloning site

Transcription

Translation

(b)

ribosome binding site

translational spacer element

cloning site

start codon

Target Gene

EC-PCR

stop codon

cloning site

Cloned into overexpression plasmid

FIGURE 4.6: *Addition of promoter and ribosome-binding site (RBS) sequences to PCR products. **(a)** Expression-PCR. A promoter and untranslated leader sequence that contains an RBS and a suitable sequence for initiation of translation is added to the PCR primer as shown. The PCR product can be transcribed into RNA using the appropriate RNA polymerase (e.g. T7) and the RNA is translated into protein in vitro. **(b)** Expression-cassette PCR. Translational elements (RBS and translational spacer element) are added to the end of the PCR primers as shown and the product (expression cassette) is then cloned into an overexpression plasmid that provides a strong regulated promoter and an efficient transcription terminator.*

to some extent by using PCR to amplify long oligonucleotides from a crude synthesis reaction without purification (see Section 4.3.2). It is also possible to fuse universal promoter and ribosome-binding site (RBS) sequences to any gene of interest by combining E-PCR or EC-PCR with other methods of PCR gene fusion (e.g. overlap extension as discussed in Section 4.3.1). These methods are also compatible with rapid PCR-based site-directed mutagenesis (see Chapter 6).

These approaches save considerable time, since the gene of interest does not need to be cloned into a vector. Because of the ease of PCR, it is simple to experiment with many genes and promoters. Also, by the incorporation of different promoters at either end of the PCR product it is possible to produce 'sense' or 'antisense' RNA transcripts by using the respective RNA polymerases. These transcripts may then be used as hybridization probes. The other main use of this type of PCR product modification is for transcript sequencing, as discussed later in Section 7.6.

4.3 Joining overlapping PCR products

4.3.1 Chimeric genes and gene construction from genomic sequence

The ease and the flexibility of PCR, compared to conventional recombinant DNA techniques, has led to it becoming a popular method for engineering DNA molecules. PCR products can be modified readily by addition of 5′ sequences to the PCR primers, as described in Section 4.1. This technique can also be adapted for gene splicing to allow the construction of chimeric genes.

Several groups have reported a more specific and universal method of construction of chimeric genes by PCR (for examples see references 12–15). The technique is known as PCR gene fusion, recombinant PCR, or splicing by overlap extension (SOE). This approach can recombine DNA molecules at precise junctions without the use of restriction endonucleases and DNA ligase. This is irrespective of the DNA sequences at the junctions. Fragments from the genes that are to be fused are generated in separate PCRs. This is shown schematically in *Figure 4.7*, with a more detailed example of the design of SOE primers being shown in *Figure 4.8*. The primers are designed so that the ends of the PCR products contain complementary sequences and so, when mixed, the strands having the matching sequences overlap and can be extended by a DNA polymerase. If using *Taq* DNA polymerase which adds an A residue to the 3′ ends of either strand of the PCR product, the primers are also designed to allow for this in generating complementary overlaps (*Figure 4.8*). The combined and extended PCR products result in a 'spliced', or fused, chimeric gene. It is also possible to perform site-directed mutagenesis (see Chapter 6) at the same time as recombination. Because of the speed, simplicity and versatility of PCR, these splicing techniques make sophisticated genetic operations possible. The method also

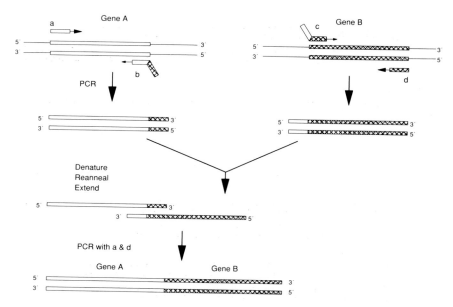

FIGURE 4.7: *PCR gene fusion. Segments of two unrelated genes (A and B) are to be fused. Separate PCRs are performed with primers a and b and c and d. Primers b and c match their template genes in their 3' portions, but their 5' portions are designed so that they are complementary to the other gene sequence. After PCR the two amplicons are mixed along with primers a and d, denatured, reannealed and primer extended with DNA polymerase in the PCR. The end of each amplicon is complementary to the other and therefore they can act as primers for each other and be extended to form the full-length recombinant chimeric gene which is amplified with primers a and d in the PCR. A modification of this method involves only three primers: a; the bridging primer, b; and d. The two DNA fragments to be fused are mixed with all three primers in the PCR. The concentration of primer b is 20- to 100-fold lower than a and d. Initially, a and b will amplify gene A and then primer b will be consumed as its concentration is low. The amplified strand of gene A, which contains a sequence complementary to the 5' end of gene B at its 3' end, can act as a primer for gene B. PCR with a and d results in the fusion of the two genes.*

avoids laborious and multiple-step molecular cloning procedures, which may be problematic, particularly when no suitable cloning sites are available.

The use of class IIS restriction enzymes for the efficient cloning of any given DNA sequence into any desired location in the absence of naturally occurring restriction sites has been described [16]. This technique employs PCR amplimers which contain an *Eam*1104 I

(a)   5′. TCACCTTATAT**GTCTATGTCTAACCGA**GCATAATCTATCGCAGCATATCAGCATTATG3′
      3′. AGTGGAATATAC**AGATACAGATTGGCT**CGTATTAGATAGCGTCGTATAGTCGTAATAC5′

                 5′CATAGACTCGCGCTCGTTAGCTGCTAGATGCTT**GATGTCCAGCTGACTCGA**TATAGCTC .3′
                 3′GTATCTGAGCGCGAGCAATCGAGGATCTACGAA**CTACAGGTCGACTGAGC**TATATCGAG .5′

(b)   5′GTCTATGTCTAACCGA-GATGTCCAGCTGACTCGA3′

      3′AGATACAGATTGGCT-CTACAGGTCGACTGAGC5′ (= 5′CGAGTCAGCTGGACATC-TCGGTTAGACATAGA3′)

(c)   5′. TCACCTTATATGTCTATGTCTAACCGAGCATAATCTATCGCAGCATATCAGCATTATG3′
      3′← ← ← ← ← ← ← ← ← ← ← ← AGATACAGATTGGCT-CTACAGGTCGACTGAGC5′

                        5′GTCTATGTCTAACCGA-GATGTCCAGCTGACTCGA → → → → → → → →3′
                        3′GTATCTGAGCGCGAGCAATCGAGGATCTACGAA**CTACAGGTCGACTGAGC**TATATCGAG →5′

(d)   5′←TCACCTTATATGTCTATGTCTAACCGAGATGTCCAGCTGACTCGA3′
      3′←AGTGGAATATACAGATACAGATTGGCTCTACAGGTCGACTGAGC5′

                 5′GTCTATGTCTAACCGAGATGTCCAGCTGACTCGATATAGCTC →3′
                 3′ACAGATACAGATTGGCTCTACAGGTCGACTGAGCTATATCGAG →5′

(e)   5′←TCACCTTATATGTCTATGTCTAACCGAGATGTCCAGCTGACTCGA → → → → → → → →3′
      3′←← ← ← ← ← ← ←ACAGATACAGATTGGCTCTACAGGTCGACTGAGCTATATCGAG →5′

                 5′GTCTATGTCTAACCGAGATGTCCAGCTGACTCGATATAGCTC →3′
                 3′←ACTGGTATATACAGATACAGATTGGCTCTACAGGTCGACTGAGC5′

FIGURE 4.8: *The design of PCR primers for SOE using* Taq *DNA polymerase.* **(a)** *Two DNA fragments (only 3′ and 5′ ends of respective fragments shown); bold type indicates the parts of each sequence to be fused; the underlined bases show the sequence from which primers are designed; note that the 5′ end of the primer sequence precedes a T residue which compensates for the 3′ A addition by* Taq *DNA polymerase.* **(b)** *The primers; the hyphen denotes the junction of the sequences from each DNA fragment and the point of fusion.* **(c)** *Priming of each target with the respective primer. Arrows indicate primer/template polymerase extension.* **(d)** *The respective ends of each PCR product, showing the incorporated primer; note the 3′ terminal A residues.* **(e)** *The annealing of the two PCR products; note that the additional A residues are complementary to the template derived from the other PCR product. Arrows indicate primer/template polymerase extension.*

recognition site (5′-CTCTTC) (*Figure 4.9*) at their 5′ ends and these are used for PCR of insert and vector (*Figure 4.10*). A minimum of two nucleotides 5′ to the CTCTTC sequence are also required for efficient cleavage. Type IIS restriction endonucleases are able to cleave several bases downstream from their recognition site and so additional terminal sequences are removed and 5′ overhangs are generated that are defined by the nucleotides present with the cleavage sites. The generation of unique, nonpalindromic sticky ends permits the formation of seamless junctions in a directional fashion. If sites for the *Eam*1104 I are present in the insert or vector then potentially this

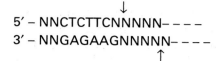

↓
5′ – NNCTCTTCNNNNN– – – –
3′ – NNGAGAAGNNNNN– – – –
                    ↑

FIGURE 4.9: *Recognition sequence 5′-CTCTTC and cleavage site of* Eam*1104 I (shown with arrows). Cleavage with* Eam*1104 I generates termini with three-nucleotide-long 5′ overhangs.*

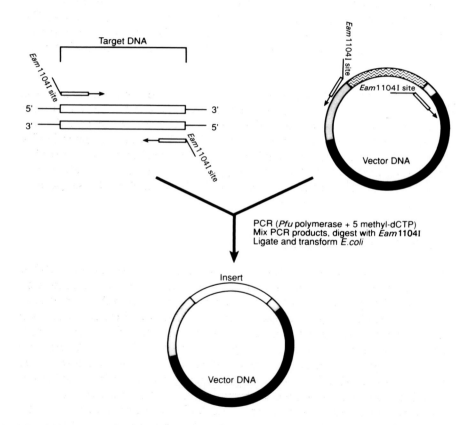

Figure 4.10: *Domain substitution by seamless cloning using the type IIS restriction endonuclease* Eam*1104 I. The domain indicated above is synthesized by PCR. The vector is also amplified except the domain. Both PCRs use amplimers with* Eam*1104 I sites at their 5′ends. Upon digestion of the PCR fragments sticky ends are generated that can be ligated. This results in the domain of interest being introduced without the addition of extra nucleotides or restriction sites.*

could be problematic. However, *Eam*1104 I activity is inhibited when dCTP in the recognition sequence is replaced with 5-methyldeoxycytosine (^{M5}dCTP). Incorporation of ^{M5}dCTP into the PCR amplicon results in a PCR product that is resistant to cleavage at the naturally occurring *Eam*1104 I sites, but the terminal amplimer encoded sites are still digestible. This method can be used for domain swapping and expression cloning. Reagents for performing this are available commercially (Seamless™ Cloning Kit, Stratagene).

Multiple gene fusions in which several bridge primers are used are also feasible, and the method can be applied to the preparation of deletion mutants or site-directed mutants. Indeed, it should be possible to construct a cDNA sequence by such techniques using the genomic DNA sequence as the template. This would potentially be useful in cases where the cDNA is very large and/or the mRNA very rare.

Finally, the ability to perform *in vitro* transcription/translation (see Section 4.2.1) on PCR products should make it possible to circumvent the heterologous expression of recombinant proteins from chimeric fusion genes. If the protein(s) of interest can be assayed specifically and with sufficient sensitivity, *in vitro* synthesis may provide sufficient material to allow screening for the desired characteristics, and then the optimum construct could be cloned and the protein expressed in greater amounts.

4.3.2 Synthetic genes

Previously published methods for *in vitro* synthesis of dsDNA molecules involved using synthetic oligonucleotides in a multistep process. It is possible to use a similar method of assembling genes in which synthetic oligonucleotides are annealed and ligated then, as a final step, amplified with the terminal oligonucleotides used as PCR primers to provide the complete gene in high yield [17]. A one-step process to perform gene synthesis using fewer oligonucleotides has been reported [18]. Four adjacent oligonucleotides having overlaps of 15–17 nucleotides are used as primers in a PCR mixture. The two outer primers are approximately 100-fold more abundant than the inner ones and this causes an asymmetric amplification of the two fragments of the total sequence. These fragments, which overlap each other, yield a double-stranded, full-length product in subsequent cycles of PCR. Direct construction of synthetic genes has been reported by several groups applying PCR to crude oligonucleotide mixtures made on automated DNA synthesizers [19–21]. This has been achieved by either synthesizing the entire top and bottom strands of the gene and then using PCR with two flanking primers, or by synthesizing one of the strands of the synthetic gene followed by PCR with the two flanking primers. Genes have been constructed up to 655 bp in length using this type of approach [17]. It should be possible to assemble longer genes by making segments of about 500 bp directly by this method and then ligating them together. The PCR amplification is essential as it selects for the full-length strand in the crude mixture of oligonucleotides (which comprises mostly shorter, failure-sequence oligonucleotides) and amplifies only full-length strands to give the target gene. Gene construction is a powerful tool in molecular biology and is particularly useful when a gene has to be synthesized with a specific codon usage for expression studies or optimization. It is also possible to assemble novel sequences, for example sequences that will act as a purification or detection tag when fused to a protein of interest.

References

1. Scharf, S.J., Horn, G.T. and Erlich, H.A. (1986) *Science,* **233,** 1076.
2. Clarke, J.M. (1988) *Nucleic Acids Res.,* **16,** 9677.
3. Marchuk, D., Drumm, M., Saulino, A. and Collins, F. S. (1991) *Nucleic Acids Res.,* **19,** 1154.
4. Kaufman, D.L. and Evans, G.A. (1990) *BioTechniques,* **9,** 304.
5. Aslanidis, C. and deJong, P.J. (1990) *Nucleic Acids Res.,* **18,** 6069.
6. Haun, R.S., Serventi, I.M. and Moss, J. (1992) *BioTechniques,* **13,** 515.
7. Haun, R.S. and Moss, J. (1992) *Gene,* **112,** 37.
8. Nisson, P.E., Rashtchian, A. and Watkins, P. C. (1991) *PCR Methods Appl.* **1,** 120.
9. Browning, K.S. (1989) *Amplifications,* **3,** 14.
10. Kain, K.C., Orlandi, P.A. and Lanar, D.E. (1991) *BioTechniques,* **10,** 366.
11. MacFerrin, K.D., Terranova, M.P., Schreiber, S.L. and Verdine, G.L. (1990) *Proc. Natl Acad. Sci. USA,* **87,** 1937.
12. Horton, R.M., Hunt, H.D., Ho, S.N., Pullen, J.K. and Pease, L.R. (1989) *Gene,* **77,** 61.
13. Yon, J. and Fried, M. (1989) *Nucleic Acids Res.,* **17,** 4895.
14. Daugherty, B.L., DeMartino, J.A., Law, M.-F., Kawka, D.W., Singer, I.I. and Mark, G.E. (1991) *Nucleic Acids Res.,* **19,** 2471.
15. Cao, Y. (1990) *Technique,* **2,** 109.
16. Padgett, K.A. and Sorge, J.A. (1996) *Gene,* **168,** 31.
17. Jayaraman, K., Shah, J. and Fyles, J. (1989) *Nucleic Acids Res.,* **17,** 4403.
18. Sandhu, G.S., Aleff, R.A. and Kline, B.C. (1992) *BioTechniques,* **12,** 14.
19. Barnett, R.W. and Erfle, H. (1990) *Nucleic Acids Res.,* **18,** 3094.
20. Ciccarelli, R.B., Gunyuzlu, P., Huang, J., Scott, C. and Oakes, F.T. (1991) *Nucleic Acids Res.,* **19,** 6007.
21. Michaels, M.L., Hsiao, H.M. and Miller, J.H. (1992) *BioTechniques,* **12,** 45.

5 Isolation and Construction of DNA Clones

Conventional methods for the isolation and characterization of cDNA clones rely on the preparation and screening of cDNA libraries. Such methods require good-quality libraries, which are difficult to prepare and take time to screen. PCR can be used to selectively amplify a specific cDNA directly from the library if the sequence is known. A modified 'anchored' PCR method can be used if only a small amount of sequence information is known. This chapter describes the methods available for cDNA isolation by these alternative PCR-based methods and RNA (RT) PCR.

5.1 PCR from libraries

PCR can be used to amplify specific DNA fragments directly from cDNA or genomic libraries (in bacteriophage lambda, cosmid or YAC vectors) if the DNA sequence of the gene of interest is already known [1]. This can be done by simply taking a small aliquot of the library, denaturing it by heating to 100°C for several minutes and then using this directly in a PCR reaction. This type of approach is also useful for direct clone characterization. For example, when screening cDNA libraries either in lambda or plasmid vectors it is common to have large numbers of positive clones to characterize. PCR can be performed directly on bacterial plaques or colonies and can allow the presence and the size of inserts to be determined rapidly by amplification with vector-specific flanking primers [2,3]. The orientation of the inserts can be determined by including a single internal primer from the gene of interest in the PCR. If the sequence is unknown, individual clones may be characterized further as described in Section 8.3.

In addition, PCR techniques are available for the isolation of novel sequences from libraries by using either the amino acid sequence of a protein or homology with other similar, known genes to derive primer sequence [4]. For instance, a species-specific enzyme already known to be conserved between other species may have its gene PCR amplified by back-translating the DNA coding sequence from a consensus amino acid sequence. Primers specific to the gene of interest are then used in conjunction with universal primers derived from a region of vector sequence next to the cloning site. For example, for a cDNA library prepared in lambda gt11, a gene-specific primer could be used along with a lambda gt11 primer from the *lacZ* gene flanking the *Eco*RI cloning site. The efficiency of this approach is greatly increased by the use of nested primers.

5.2 Anchored PCR

'Anchored PCR' is a method that involves adding the template for a universal primer to one end of the DNA fragment to be amplified. Several modifications of this type of approach have been described and the use of 'vectorette PCR' on genomic DNA is described later (see Section 8.1). In this section we describe two anchored PCR methods for cDNA isolation: rapid amplification of cDNA ends (RACE)-PCR and ligation-anchored PCR.

5.2.1 RACE-PCR

The RACE procedure [5–7] or one-sided PCR [8], illustrated in *Figures 5.1* and *5.2*, is a method by which the PCR technique can be used to amplify the 3' and 5' ends of a cDNA using a small stretch of known sequence within the gene. This small stretch can be derived by back-translation from the amino acid sequence of the protein (either from homology with other related proteins or determined experimentally) or can be from DNA sequence homology with other members of the gene family. The procedure can be made even more powerful by the use of nested primers, which reduces nonspecific amplification and ensures the production of relatively pure specific product. The 3' and 5' ends of cDNA are isolated in separate reactions. The isolation of the 5' end is especially problematic for large gene transcripts and those of low abundance. The standard procedure of rescreening libraries is time-consuming and may not result in 5'-end clones; RACE-PCR provides a useful alternative.

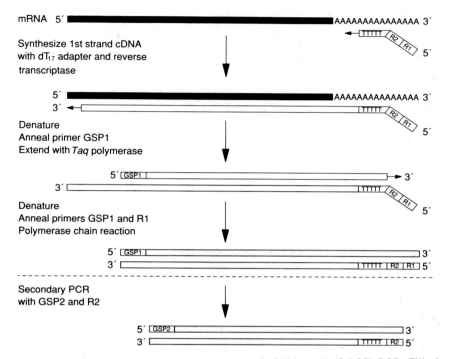

FIGURE 5.1: *3' End rapid amplification of cDNA ends (RACE)-PCR. Filled rectangles represent mRNA, open rectangles represent DNA strands being synthesized, and stippled rectangles represent DNA previously synthesized. TTTTTR2R1 is the dT$_{17}$ adapter. The diagram is simplified to illustrate only how the new product formed during the prior step is utilized. R1 and GSP1 are RACE anchor and gene-specific primers, respectively; R2 and GSP2 are nested RACE anchor and nested gene-specific primers, respectively. Secondary PCR with R2 and GSP2 is optional, depending on the abundance of the specific mRNA in the mRNA preparation used.*

For 3' RACE a reverse transcription primer that contains an oligo dT sequence linked to an adapter sequence (which may be long enough to permit binding of two nested primers or shorter for single primer binding) is used to prime the first-strand cDNA synthesis. Primary amplification is then performed with a gene-specific primer (GSP1) and the outer primer R1. A small fraction of the first amplification is then used for secondary amplification with a nested gene-specific primer (GSP2) and inner primer R2 (see *Figure 5.1*).

For 5' RACE, cDNA is prepared with either a specific primer or random hexamers and then the newly synthesized cDNA strand has

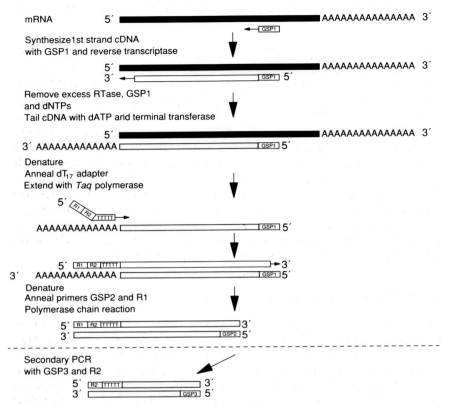

FIGURE 5.2: *5' End RACE-PCR. Filled rectangles represent mRNA, open rectangles represent DNA strands actively being synthesized, and stippled rectangles represent DNA previously synthesized. R1R2TTTTT is the dT₁₇ adapter. R1 and GSP1 are RACE anchor and gene-specific primers, respectively; R2, GSP2 and GSP3 are nested RACE anchor and nested gene-specific primers, respectively. GSP1 can be used for cDNA synthesis and for primary PCR but increased specificity is obtained by using GSP2 for the PCR. Secondary PCR with R2 and GSP3 is optional, depending on the yield and specificity of the primary products.*

a homopolymeric tail (e.g. A residues) added to its 3' end using terminal transferase. The complementary reverse transcription anchor primer is then used to generate second-strand cDNA. This dsDNA can then serve as the template for PCR reactions as described for 3' RACE.

To perform 5' and 3' RACE-PCR, either the components can be purchased from many commercial suppliers (e.g. cDNA kits;

Amplitaq®, etc.) and the primers synthesized by the researcher, or 5′ and 3′ RACE-PCR kits are available commercially (Gibco-BRL, CLONTECH).

Other variations of RACE-PCR have been reported and most of these differ in the method of adapter ligation. One such technique is Marathon™ cDNA amplification for 5′ and 3′ RACE (CLONTECH) which uses adapters as shown. The primer site is not present on the adapter- ligated cDNAs but is created by extension of the gene-specific primer in the first cycle which greatly enhances the specificity of the PCR. The adapter primers are also shorter than the adapter itself and so any cDNAs which have the full-length adapter on both ends (which may occur, for example, due to incomplete amine addition during adapter synthesis) will form an intramolecular structure (panhandle) due to annealing of the complementary adapter ends. During thermal cycling this structure is favored over intermolecular annealing of the adapter primer so little full-length amplification occurs across regions which have adapter on both ends.

5.2.2 Ligation-anchored PCR

This is a simple, efficient and sensitive technique for the amplification of cDNAs where the 5′ end is of unknown sequence [9, 10]. In this method, T4 RNA ligase is used to covalently link the 'anchor' oligonucleotide to first-strand cDNAs. Amplification is then carried out with one primer specific for a sequence within the gene of interest and one primer specific for the anchor (see *Figure 5.3*). The anchor oligonucleotide must be 5′ phosphorylated, this is necessary for ligation to the cDNA. The 3′ end of the anchor oligonucleotide is blocked, for example, by addition of a dideoxynucleotide using terminal deoxynucleotidyl transferase or by the incorporation of an amino group during oligonucleotide synthesis. This 3′ modification is required to prevent the ligation of more than one anchor to the cDNA. A modified version of this procedure is available commercially as a kit (3′ AmpliFINDER™ RACE, CLONTECH Laboratories).

5.2.3 RNA ligase-mediated RACE

RNA ligase-mediated RACE (RLM-RACE) [11], also known as reverse ligation-mediated PCR (RLPCR) [12], can allow the 5′ ends of cDNAs to be cloned. An RNA oligonucleotide anchor sequence, either a synthetic oligo or a run-off transcript, is ligated to the 5′ ends of the mRNA. The mRNA is treated chemically [11] or enzymatically [13] to remove the 7-MeGppp cap which is covalently attached to the 5′ end of all mRNA transcripts. In the enzymatic method, alkaline

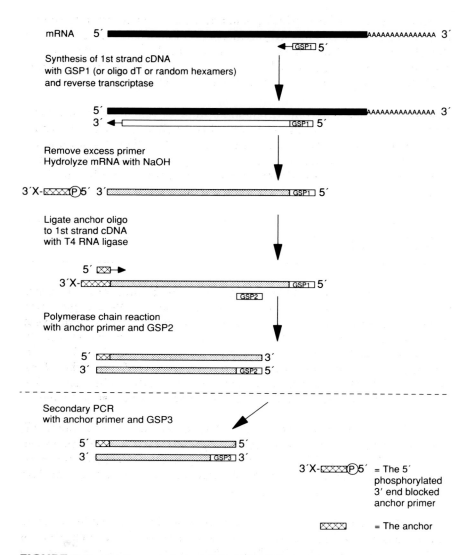

FIGURE 5.3: *Ligation-anchored (LA) PCR. Filled rectangles represent mRNA, open rectangles represent DNA strands actively being synthesized, and stippled rectangles represent DNA previously synthesized. The 5'-phosphorylated, 3'-blocked anchor oligonucleotide is ligated to the 3' end of the first-strand cDNA with T4 RNA ligase. PCR is performed with oligonucleotide primers specific for a sequence within the anchor and for a sequence within the cDNA/gene sequence (GSP2). Secondary PCR with the anchor primer and GSP3 is optional.*

phosphatase or calf intestinal phosphatase is used to remove the 5'-terminal phosphates from any degraded RNAs. After inactivation of the phosphatase, the full-length mRNAs are decapped with tobacco acid pyrophosphatase, which leaves a 5' phosphate on decapped mRNAs, rendering them substrates for T4 RNA ligase.

The anchor-modified RNA is reverse transcribed to generate first-strand cDNA, which is then amplified by PCR using a 3'-end gene-specific amplimer and an anchor amplimer (DNA). This technique can also be used for 3' RACE since the RNA oligonucleotides also ligate to the 3' end of cytoplasmic RNAs. In this case a gene-specific amplimer going towards the 3' end of the gene is used.

5.3 RNA (RT) PCR

Amplification of RNA by PCR can be performed by annealing a primer to the RNA template and then synthesizing a cDNA copy using reverse transcriptase (RT), followed by PCR. Some DNA polymerases can be used for this step, for example the thermostable *T. thermophilus* (*Tth*) DNA polymerase (see Section 2.2.4) in the presence of manganese can reverse transcribe RNA. Therefore, because *Tth* DNA polymerase can utilize both DNA and RNA templates, the whole procedure can be carried out in a single tube.

For mRNA that possesses a poly(A) tract at the 3' end, oligo dT, random hexamers or a gene-specific primer can be used to prime cDNA synthesis. Viral RNA templates (e.g. retroviruses, rhinoviruses, etc.) or nonpolyadenylated RNA can be copied using random hexamers or specific target primers.

RNA (RT) PCR is a highly sensitive tool in the study of gene expression at the RNA level and, in particular, in the quantitation of mRNA or viral RNA levels. This technique, also known as message amplification phenotyping (MAPPing), permits the simultaneous analysis of a large number of mRNAs from small numbers of cells [14]. RNA PCR can also be used as a first step in preparing a cDNA library by PCR of all of the mRNAs in a sample of cellular RNA. Such methods have been successful and, with continuous improvements being reported, it may soon be possible to construct a cDNA library from small numbers of cells or even single cells. This would be important in situations where only small numbers of cells are available or when the cells of interest cannot be propagated.

Reverse transcriptase is usually used to synthesize first-strand cDNA from RNA. Reverse transcriptases can be purified from several sources (e.g. avian myeloblastosis virus (AMV) and Moloney murine leukemia virus (MMLV)). AMV reverse transcriptase is an RNA-dependent DNA polymerase that uses single-stranded RNA as a template and can synthesize a complementary DNA (cDNA) in the 5′ to 3′ direction if a primer is present. As well as DNA polymerase activity this enzyme also exhibits ribonuclease H activity, which is specific for RNA : DNA heteroduplex molecules.

MMLV reverse transcriptase acts in the same way as AMV reverse transcriptase; however, it lacks DNA endonuclease activity and has lower RNase H activity. Thus it has a greater chance of producing full-length copies of large mRNA species. There is also an RNase H⁻ MMLV reverse transcriptase (Superscript Plus™, Gibco-BRL) which is a modified form with point mutations in the RNase H coding sequence of a cloned MMLV reverse transcriptase gene. This form of the enzyme eliminates degradation of the RNA molecule during first-strand synthesis, resulting in more full-length synthesis and greater yields of cDNA.

5.4 PCR-based cDNA library construction

cDNA libraries are normally constructed directly with RNA from the organism or tissue of interest. Researchers who work on homogeneous populations of cultured cells have potentially large amounts of starting material, and usually a library of 10^5 recombinants is sufficient to be representative of the mRNA population. For many plant and animal tissues this also presents little problem as there is sufficient starting material and perhaps only a few cell types. For example, with cultured cells or tissues which are relatively large and have a low degree of cellular heterogeneity it should be possible to extract more than an adequate amount of RNA for library preparation by standard methods. Although the yield of RNA varies with cell type and tissue, as a general rule it should be possible to extract 0.5–30 mg total RNA from approximately 1 g of tissue or approximately 5×10^8 cultured cells. Polyadenylated (poly(A)⁺) RNA normally constitutes about 1–2% of the total RNA population. Most eukaryotic mRNAs are polyadenylated at their 3′ ends and this region can be used to prime first-strand cDNA synthesis by using a complementary primer, oligo(dT), with the enzyme reverse transcriptase. Second-strand synthesis is then followed by cloning into a suitable vector (bacteriophage lambda or plasmid-based). However, there are many instances where only small amounts of tissue are available or the cells of interest are difficult to propagate or, indeed, cannot be propagated (e.g. endoparasites, neural cells and from some specific anatomical regions). There are other examples where small amounts of starting material are used for preparing libraries, for example in subtraction cloning. In this method mRNAs that are not cell-type specific are removed, leaving small amounts that can be reverse transcribed into cDNA that is specific to the cell type. There have been several published procedures, both theoretical and practical, where an entire population of cDNAs has been amplified and libraries prepared [5,15]. The main problems with PCR-based cDNA library construction are:

(i) attaching sequences that can be used with common primers for all cDNA molecules; and
(ii) misrepresentation or loss of specific sequences.

Poly(A)$^+$ mRNAs have a common sequence at their 3′ ends that can be converted into a sequence of T residues when first-strand cDNA is synthesized using oligo(dT). The addition of a predetermined 'anchor' sequence to the 5′ end of the oligo(dT) provides a 'handle' for one end of the resulting cDNAs. Restriction enzyme recognition sites can also be incorporated into this 'anchor' (see Section 4.1.1) to allow subsequent manipulations. There is a variety of ways to attach a 'handle' to the other end of each cDNA. Homopolymer tailing (usually G residues) with the enzyme terminal transferase has been used for specific cDNA amplification of mixed populations of oligo(dT) primed cDNAs. As described above, a specific sequence can be provided next to the homopolymer tail to provide another 'anchor'. Other techniques, such as ligation of double-stranded oligonucleotides to both ends of blunt-ended cDNA molecules, amplification of cDNAs that have been ligated into plasmids, ligation of single-stranded 'anchor' primer oligonucleotides using T4 RNA ligase or blunt-ended vectorette ligation (see Section 8.1), could all potentially be applied to PCR-based cDNA library construction.

Misrepresentation or loss of cDNAs during PCR-based library construction is a major concern; however, steps can be taken to minimize the problems. There are several likely causes of misrepresentation. Certain cDNA sequences may be difficult to amplify due to amplification refractory sequences which halt the DNA polymerase or which may be more difficult to denature than other sequences. Also, short sequences are more efficiently reverse transcribed and amplified than longer ones. Most mRNAs are less than 3 kb in length; however, there are relatively rare mRNAs longer than this. Specific PCR products longer than 10 kbp have been reported and so it should be possible in principle to PCR amplify the majority of mRNAs/cDNAs in any given population.

Microscale methods of RNA preparation or the use of crude boiled cell supernatants can be applied to cDNA library construction.

5.5 PCR with degenerate primers

It is possible to amplify an uncloned gene by using protein sequence data to design degenerate primers for PCR. Such primers have been used to isolate members of multigene families and to isolate genes across species. Degenerate primers can also be used in 'anchored' PCR methods, as described above. Amplified DNA from the gene of interest can then be used as a probe to screen relevant cDNA or genomic libraries, Southern or Northern blots at high stringency. The design of degenerate primers by back-translation from peptide sequences is complicated by the fact that most amino acids are encoded by more than one codon. Some amino acids have only one or two possible codons (Met, Trp, Cys, Asp, Glu, Phe, His, Lys, Asn, Gln, Tyr) but others have up to six possible codons (there are six possibilities for Leu, Arg and Ser). If possible, degenerate primers should be designed from protein such that there is a bias towards amino acids encoded by only one or two codons. It may also be possible to design primers so that the preferential codon usage in different species is maintained [16]. As an alternative to the design of highly degenerate primers where there are many positions of three- or fourfold base redundancy it is possible to incorporate the universal base inosine at these positions, which reduces the complexity of the mixture of primer sequences [17]. Inosine is a rare natural purine base which can base pair with the four bases A, C, G and T.

Deoxyinosine is the most widely used universal base which can base pair with A, C, G or T but with different affinities. The universal nucleotide phosphoramidite [18] (Glen Research) and the universal base phosphoramidite [19] (CLONTECH) introduce modified bases into primers that can pair equally well with all four natural bases. These have PCR applications which include cDNA cloning based on back-translation of amino acid sequence, cloning members of gene families and cross-species cDNA isolation.

The three bases at the 3′ end of a degenerate primer must be perfectly matched with the template and should not include inosine. Beyond this, mismatches can be tolerated although it is important to keep the mismatched bases at the 3′ end to a minimum. PCR primers with a potential redundancy of 73 728-fold have been used successfully to amplify DNA. Reaction conditions have to be determined experimentally to define those that will achieve maximum amplification of unique and specific products. Multiple sequence alignments of highly conserved regions of proteins have been used to generate 'consensus' primers capable of amplifying corresponding coding regions from the same gene from diverse species or for amplifying other members of multigene families.

5.6 Differential display PCR

Arbitrarily primed PCR can use an arbitrarily selected primer or two such primers to produce PCR fingerprints of complex DNA or RNA (see Section 9.1). At a sufficiently low temperature the arbitrary primer (with no degeneracies) anneals to the best matches in the template and the newly synthesized strand becomes available for another priming event with the same or a different primer. Products are then further amplified by PCR; with some primers no differences will be seen in the range of products derived from different templates but with other primers there will be noticeable differences. Polymorphic differences in band patterns on a gel can be used for genetic mapping, phylogenetics and epidemiology. Similarly, arbitrarily primed PCR of RNA [20–22] can be used to detect and clone mRNA transcripts differing in abundance and reflecting differential gene expression between tissues or cell lines, etc. This technique, also known as differential display reverse transcription PCR (DDRT-PCR) employs arbitrary 10mer primers together with anchored oligo(dT) primers for generating random amplicons using first-strand cDNA as template. A large set of radioactive amplicons can be obtained by the use of different primer combinations and a radioactive nucleotide in the reactions, these can be resolved (e.g. on DNA sequencing gels). Bands derived from differentially expressed genes can then be detected by autoradiography if required, these can be eluted from the gel and amplified in a second PCR for subsequent cloning.

References

1. Friedman, K.D., Rosen, N.L., Newman, P.J. and Montgomery, R.R. (1988) *Nucleic Acids Res.,* **16,** 8718.
2. Gussow, D. and Clackson, T. (1989) *Nucleic Acids Res.,* **17,** 4000.
3. Saiki, R.K., Gelfand, D.H., Stoffel, S., Scharf, S., Higuchi, R., Horn, G.T., Mullis, K. and Erlich, H. (1988) *Science,* **239,** 487.

4. Jansen, R., Kalousek, F., Fenton, W.A., Rosenberg, L.E. and Ledley, F.D. (1989) *Genomics,* **4,** 198.

5. Belyavsky, A., Vinogradova, T. and Rajewsky, K. (1989) *Nucleic Acids Res.,* **17,** 2919.

6. Frohman, M.A., Dush, M.K. and Martin, G.R. (1988) *Proc. Natl Acad. Sci. USA,* **85,** 8998.

7. Frohman, M.A. and Martin, G.R. (1989) *Technique,* **1,** 165.

8. Ohara, O., Dorit, R.L. and Gilbert, W. (1989) *Proc. Natl Acad. Sci. USA,* **86,** 5673.

9. Dumas, J.B., Edwards, M., Delort, J. and Mallet, J. (1991) *Nucleic Acids Res.,* **19,** 5227.

10. Troutt, A.B., McHeyzer-Williams, M.G., Pulendran, B. and Nossal, G.J.V. (1992) *Proc. Natl Acad. Sci. USA,* **89,** 9823.

11. Gorovsky, M.A. and Liu, X. (1993) *Nucleic Acids Res.,* **21,** 4954.

12. Fromont-Racine, M., Bertrand, E., Pictet, R. and Grange, T. (1993) *Nucleic Acids Res.,* **21,** 1683.

13. Mandl, C.W., Heinz, F.X., Puchhammer-Stockl, E. and Kunz, C. (1991) *BioTechniques,* **10,** 484.

14. Brenner, C.A., Tam, A.W., Nelson, P.A., Engelman, E.G., Suzuki, N., Fry, K.E. and Larrick, J.W. (1989) *BioTechniques,* **7,** 1096.

15. Tam, A.W., Smith, M.M., Fry, K.E. and Larrick, J.W. (1989) *Nucleic Acids Res.,* **17,** 1269.

16. Grantham, R., Gantier, C., Gouy, M., Jacobzone, M. and Mercier, R. (1981) *Nucleic Acids Res.,* **9,** 43.

17. Ehlen, T. and Dubeau, L. (1989) *Biochem. Biophys. Res. Commun.,* **160,** 441.

18. Nichols, R., Andrews, P.C., Zhang, P. and Bergstrom, D.E. (1994) *Nature,* **369,** 492.

19. Loakes, D. and Brown, D.M. (1994) *Nucleic Acids Res.,* **22,** 4039.

20. Liang, P. and Pardee, A.B. (1992) *Science,* **257,** 967.

21. Welsh, J., Chada, K., Dala, S.S., Cheng, R., Ralph, D. and McClelland, M. (1992) *Nucleic Acids Res.,* **20,** 4965.

22. Bauer, D., Müller, H., Reich, J., Riedel, H., Ahrenkiel, V., Warthoe, P. and Strauss, M. (1993) *Nucleic Acids Res.,* **21,** 4272.

6 PCR Mutagenesis

Many site-directed mutagenesis procedures involving deletion, insertion and point mutations were documented before the advent of PCR. Both single-stranded and double-stranded oligonucleotide-directed methods were developed, and all involved multiple recombinant DNA steps and were not 100% efficient. PCR has made it possible to introduce deletion, insertion and substitution mutations in a target DNA by simple and efficient procedures.

The possible uses of mutagenesis are enormous and the structure/function relationships of many proteins are currently being studied using these techniques. One of the most popular uses of these methods is in protein engineering using antibody genes (for reviews see references [1,2]).

6.1 Deletion of sequences

Deletion of defined sequences from a target gene can be achieved by several PCR-based methods. The technique illustrated in *Figure 6.1* uses a primer that spans the region which is to be deleted in combination with a second primer [3]. The mutated PCR product is then usually used for fragment swapping with the corresponding wild-type sequence. An alternative procedure uses the method of PCR gene fusion (see Section 4.3.1) to construct deletion mutants (see *Figure 6.2*), in an analogous manner to that used to construct chimeric genes [4]. Defined mutations can be made without the need for restriction endonuclease sites and fragment swapping. For dsDNA that has been cloned into a plasmid vector, it is possible to perform deletion mutagenesis by using oligonucleotide primers which are designed in inverted tail-to-tail directions to amplify the target sequence and the plasmid vector (see *Figure 6.3*) [5]. Vent™ DNA polymerase was used for this method as *Taq* DNA polymerase adds an additional nontemplated nucleotide to its PCR products and

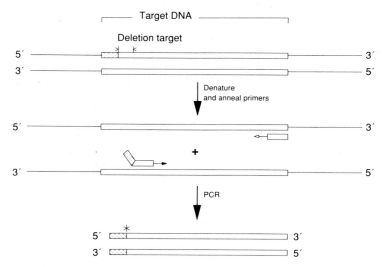

FIGURE 6.1: *Construction of a deletion mutant. The example shown is one in which the primer used for mutagenesis is at one end of the target DNA. Incorporation of convenient restriction enzyme sites in the mutagenic primer and in the amplified DNA permits replacement of a wild-type sequence with the mutated one by standard cloning methods; therefore deletions can be made at defined positions in a simple and efficient way.*

generates 3′ overhanging ends. This inverted amplification method can also be used for generating insertion and base-substitution mutations.

A novel method for mutagenesis has been described which uses high-fidelity thermostable *Pfu* DNA polymerase to generate copies of the plasmid by linear amplification (*Figure 6.4*), incorporating the mutation of interest (deletion, base substitution or insertion). Treatment with *Dpn*I endonuclease specifically degrades parental DNA since DNA isolated from most *E. coli* strains is *dam* methylated and susceptible to *Dpn*I digestion. The remaining mutated DNA is then used to transform *E. coli*, with mutagenesis efficiencies usually greater than 80%. This technique avoids the need for single-strand DNA templates, subcloning into specialized vectors or unique restriction sites. This technique has been developed by Stratagene (QUIKchange™ Site Directed Mutagenesis Kit).

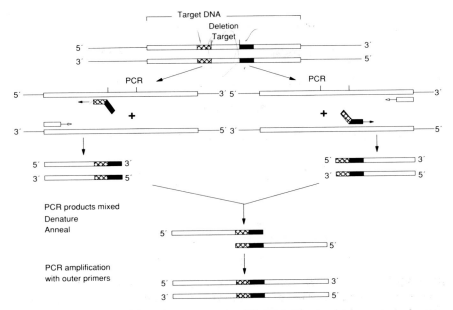

FIGURE 6.2: *Method to produce a specific deletion in a gene at any position by PCR-fusion. Two specific fragments are amplified in separate PCRs using the primers shown. The primers required for making the deletion each have a 5'sequence corresponding to the fusion flanking sequence of the opposite fragment and are complementary to each other. Aliquots of the two initial PCR reactions are mixed in a fusion PCR with the outer primers, to produce a gene with the defined deletion.*

6.2 Base substitutions

Introduction of specific base substitutions into genes can be performed using gene fusion of fragments which have had mutations introduced by use of mutagenic PCR primers (see *Figure 6.5*) [6,7]. This method is similar to one described for preparing chimeric genes (see Section 4.3.1) and deletion (see Section 6.1) or insertion mutants (see Section 6.3). It is also possible to use three primers instead of four, described above, for construction of the base-substituted DNA [8]. The first round of amplification is performed with the mutagenic primer and the antiparallel universal primer, and the second round uses the purified first PCR fragment as a primer together with the

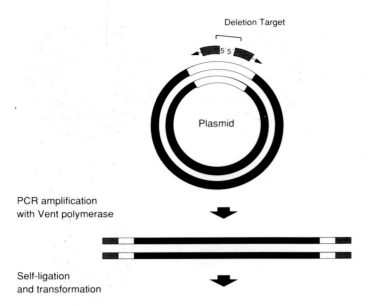

FIGURE 6.3: *Generation of deletion mutants by inverted PCR. PCR is performed in inverted directions with primers that have a gap between their 5′ ends which defines the required deletion. The PCR product is self-ligated and used to transform* E. coli. *Insertion and base-substitution mutations can also be constructed by appropriate alteration of the primers.*

second universal primer. Problems can arise due to the nontemplated addition of a single nucleotide to the PCR products, which generates a 3′ overhanging residue when *Taq* DNA polymerase is used. This can be circumvented by reducing the concentration of the dNTPs in the PCR (although this is not guaranteed to be effective), by removal of the unwanted base (using, for example, T4 DNA polymerase), or by the use of a DNA polymerase that does not exhibit this property (e.g. Vent™ DNA polymerase). Alternatively, the primers may be designed to take into account the single nucleotide addition phenomenon when *Taq* DNA polymerase is used. Since the additional residue is almost always an A residue, one solution to the problem is to prepare a mutagenic primer, designed such that the first 5′ nucleotide follows a T residue in the same strand [9] in an analogous manner to that described for generating chimeric genes (see Section 4.3.1). Therefore, whether or not an extra residue has been added to the first PCR product, the final product will have the correct sequence. Alternatively, by designing the mutagenic primer such that its 5′ end immediately follows the wobble position of a codon [10], the addition of any extra nucleotide at the 3′ end of the complementary strand will, in many cases, be tolerated, since there will be no change in the amino acid sequence encoded by the altered codon.

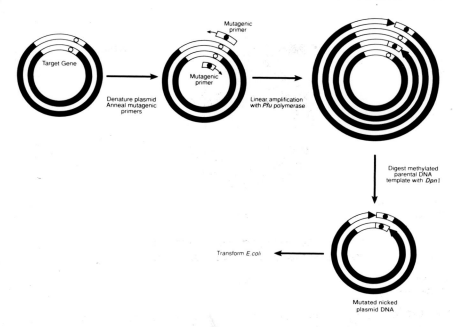

FIGURE 6.4: *Site-directed mutagenesis using* Pfu *DNA polymerase and* Dpn*I digestion. Supercoiled, double-stranded plasmid DNA is used as a template for linear amplification using* Pfu *DNA polymerase and two synthetic oligonucleotide primers, each containing the desired mutation.* Dpn*I endonuclease is used to selectively digest parental DNA which is likely to be* dam *methylated, since most* E. coli *strains methylate DNA, whereas amplified mutated DNA is resistant to digestion. The nicked vector DNA, which contains the desired mutation, is transformed into* E. coli.

Site-specific mutagenesis can be performed using asymmetric PCR with a single mutagenic primer and two flanking primers (*Figure 6.6*) [11]. This technique can be used for generating large mutations and, in this respect, is more efficient than alternative methods [6, 7] and only requires a single mutagenic primer.

Unique site elimination (USE) mutagenesis is a powerful method for performing site-directed mutagenesis in plasmid vectors [12]. This procedure has been modified and a PCR step added, which improves the overall efficiency of the technique [13]. The principle of this method is outlined in *Figure 6.7*. The advantage over other site-directed mutagenesis protocols is that this technique can be performed on any plasmid vector as long as there is a unique restriction site that can be mutated into another site and which is not present in the target gene. This avoids the need to clone the target

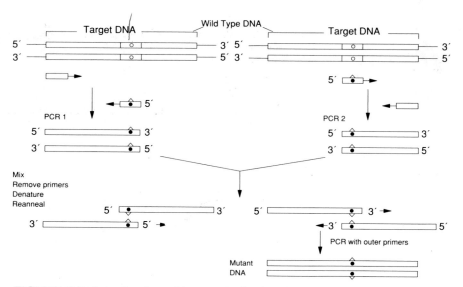

FIGURE 6.5: *Introduction of base-substitution mutations into genes by PCR overlap extension mutagenesis. The position of the base-substitution that is to be introduced is shown by a circle in the target DNA and the mutagenic primers. The empty circle denotes the wild-type sequence and the filled circle denotes the mutagenic base(s) to be introduced. Two separate PCRs result in two overlapping fragments which contain the desired mutation; these are fused and amplified in a secondary PCR.*

gene into a specific vector in order to perform the mutagenesis. Multiple rounds of mutagenesis can be performed by changing the modified restriction site back to the original by use of a selection primer coding for the original sequence, so-called 'toggle' primers. The PCR-based USE method has several advantages over the original USE method. The PCR product acts as a long primer with linked mutations for second-strand synthesis, compared to the USE method which depends on the simultaneous annealing of two primers and subsequent linkage. Also, both strands of the target DNA are substrates for second-strand synthesis, and larger PCR products will anneal more efficiently to the target DNA than will oligonucleotide primers containing the mutagenic sequences. PCR-USE is more efficient than USE (100% as compared to about 80%) and this reflects the inefficiencies in primer coupling during second-strand synthesis. By using thermostable DNA polymerases with low misincorporation rates, such as Vent™ or *Pfu*, the frequency of any unwanted mutations during PCR can be minimized.

Multiple mutations can be introduced either simultaneously or sequentially using PCR-USE (or USE). For sequentially introduced

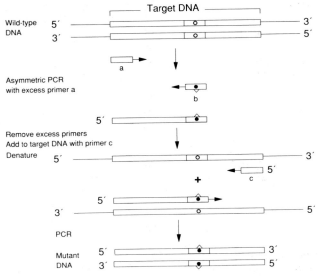

FIGURE 6.6: *Introduction of base-substitution mutations into genes by asymmetric PCR using a short mutagenic primer (b) and an excess of wild-type 5′ primer (a). The empty circle denotes the wild-type sequence, and the filled circle denotes the mutagenic base(s) to be introduced. After removal of unincorporated primers, the large single-stranded product is added to a symmetric PCR with wild-type 3′ primer (c) and a catalytic amount of wild-type template.*

mutations, it is possible to revert the introduced restriction enzyme recognition site back to the original one (e.g. *Afl*III/*Bgl*II, *Eco*RI/*Eco*RV, or *Hind*III/*Mlu*I) or to eliminate additional non-essential restriction sites.

If several mutations have to be introduced into a target gene, an alternative to multiple rounds of mutagenesis is to perform extension of several mutagenic primers by T4 DNA polymerase or Sequenase, followed by PCR amplification [14].

6.3 Insertion mutagenesis

Methods for the introduction of insertion sequences into target genes at specific places can be divided into those involving either short or long insertion sequences. For the introduction of short sequences this can be performed simply by the addition of the extra sequence at the

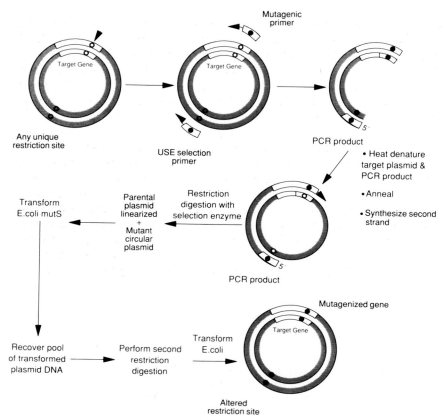

FIGURE 6.7: *PCR-based unique site elimination (USE) mutagenesis. The target gene is cloned into a plasmid vector which contains a unique restriction site. The empty circle denotes the wild-type sequence and the filled circle denotes the mutagenic base(s) to be introduced. Two oligonucleotide primers are used: one is directed to the target gene and carries the desired mutation; the other, called the USE selection primer, carries a mutation in a unique nonessential restriction site. The PCR-based USE method involves PCR using these two primers; the product contains both mutagenic oligonucleotides at its ends. This product is denatured and used as a primer for second-strand synthesis of the denatured target plasmid. The resultant DNA is digested with the selection restriction enzyme, the site of which was mutated by the incorporation of the USE primer. Therefore, plasmids containing the USE primer are resistant to digestion and nonmutated plasmids, not containing the USE primer, will be linearized. E. coli mut S, which is unable to repair mismatches in duplex DNA, is transformed with the digested plasmid DNA. Covalently closed, circular DNA is 100- to 1000-fold more efficient for transformation than linear DNA, and so the majority of transformants contain mutated plasmids. Transformants are then pooled and plasmid DNA is prepared from the mixed population. The DNA is again digested with the selection restriction enzyme and transformed into the desired E. coli host. The second digestion is required since replication of the mutated plasmid will give both wild-type and mutant progeny.*

5′ end of a PCR primer and, if a suitable restriction enzyme site has also been incorporated, the PCR product can be cloned. Short sequences can also be incorporated at defined positions using overlap extension of two overlapping fragments containing the insertion sequence (see *Figure 6.2* for analogy), whereby the additional sequences are incorporated into the 5′ ends of the primers. Each of these insertional methods is limited in the size of the insertion that may be introduced into the target gene, due to the size limitation of the sequences that can be added to the primers. It is possible to fuse longer insertion sequences by combining large insertion fragments with flanking PCR fragments by overlap extension methods; however, these are likely to be cumbersome. An alternative method for the introduction of large insertional elements by PCR has been reported [15]. This method, known as 'sticky feet' mutagenesis, can insert large regions of DNA precisely into an unrelated target sequence (*Figure 6.8*). It requires the target gene to be cloned into a single-stranded vector such as M13 or a phagemid; however, this is

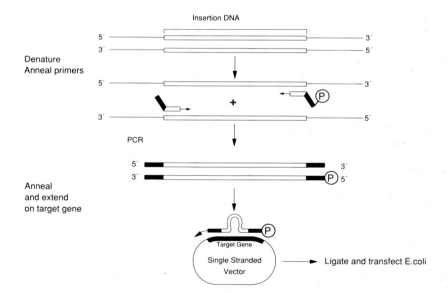

FIGURE 6.8: *'Sticky feet' directed insertional mutagenesis. The 3′ ends of the PCR primers are designed to amplify the region to be inserted into the target DNA, while the 5′ ends contain tags complementary to the target ('sticky feet'). One of the primers is phosphorylated so that ligation of the circularized DNA can occur. The single-stranded vector is grown in a* dut ung *strain of* E. coli. *After annealing and extension of the mutagenic strand, the heteroduplex is transformed into a wild-type (*ung+*) strain. The template strand is destroyed in the wild-type strain and the mutagenized strand is propagated.*

not usually a great drawback. By the use of a strain of *E. coli* (*dut ung*) that incorporates uracil in place of thymine, it is possible to have a powerful selection for mutants [16]. The single-stranded template is grown in a *dut ung* strain; after annealing and extension of the mutagenic strand, the heteroduplex is transformed into a wild-type (*ung⁺*) strain, where the template is selectively destroyed.

References

1. Winter, G. (1993) *Curr. Opin. Immunol.*, **5**, 253.
2. Skerra, A. (1993) *Curr. Opin. Immunol.*, **5**, 256.
3. Vallette, F., Mege, E., Reiss, A. and Adesnik, M. (1989) *Nucleic Acids Res.*, **17**, 723.
4. Kahn, S.M., Jiang, W., Borner, C., O'Driscoll, K. and Weinstein, I.B. (1990) *Technique*, **2**, 27.
5. Imai, Y., Matsushima, Y., Sugimara, T. and Terada, M. (1991) *Nucleic Acids Res.*, **19**, 2785.
6. Higuchi, R., Krummel, B. and Saiki, R.K. (1988) *Nucleic Acids Res.*, **16**, 7351.
7. Ho, S.N., Hunt, H.D., Horton, R.M., Pullen, J.K. and Pease, L.R. (1989) *Gene*, **77**, 51.
8. Landt, O., Grunert, H.-P. and Hahn, U. (1990) *Gene*, **96**, 125.
9. Kuipers, O.P., Boot, H.J. and deVos, W.M. (1991) *Nucleic Acids Res.*, **19**, 4558.
10. Sharrocks, A.D. and Shaw, P.E. (1992) *Nucleic Acids Res.*, **20**, 1147.
11. Perrin, S. and Gilliland, G. (1990) *Nucleic Acids Res.*, **18**, 7433.
12. Deng, W.P. and Nickoloff, J.A. (1992) *Anal. Biochem.*, **200**, 81.
13. Ray, F.A. and Nickoloff, J.A. (1992) *BioTechniques*, **13**, 342.
14. Stappert, J., Wirsching, J. and Kemler, R. (1992) *Nucleic Acids Res.*, **20**, 624.
15. Clackson, T.P. and Winter, G. (1989) *Nucleic Acids Res.*, **17**, 10163.
16. Kunkel, T.A., Roberts, J.D. and Zakour, R.A. (1988) *Methods Enzymol*, **154**, 367.

7 Sequencing PCR Products

Several approaches for DNA sequencing of PCR products permit the rapid determination of sequences of interest. It is also possible to clone PCR products into suitable vectors based on either M13 or plasmids and then sequence the resulting recombinants. Now that improved methods of cloning PCR products are available (see Chapter 4), this approach is often used. However, sequencing data by this route suffers drawbacks, for example more manipulations are required in first generating the recombinants for sequencing, and each recombinant is the product of a single DNA molecule from the PCR. Therefore, if DNA sequence information does not already exist, several recombinants should be sequenced to obtain a consensus sequence and thus eliminate anomalous sequence errors that may have been introduced by the DNA polymerase enzyme during PCR. PCR products can be sequenced using the 'dideoxy' or 'Sanger' approach [1]; alternatively, the 'chemical' or 'Maxam and Gilbert' method [2] may be used. Dideoxy sequencing is currently the method of choice for sequencing PCR products in most laboratories. Sequencing is carried out using a DNA polymerase to extend a primer along a single-stranded template in the presence of the four dNTPs (dATP, dCTP, dGTP and dTTP). Originally, the Klenow fragment of DNA polymerase I was used [1]; however, other polymerases have become popular, for example T7 DNA polymerase [3] and thermostable DNA polymerases such as *Taq*/Amplitaq® DNA polymerase [4] (P.E. Applied Biosystems), Vent™ exo⁻ (New England Biolabs), *Pfu* exo⁻ (Stratagene) or *Bst* DNA polymerase I [5] (BioRad). The DNA to be sequenced can be double- or single-stranded. If it is double-stranded DNA, then the strands have to be separated by thermal denaturation. Single-stranded DNA can be used directly in the sequencing reactions.

The sequencing reaction is terminated in a random fashion by the incorporation of a dNTP analog, a dideoxynucleoside triphosphate (ddNTP), producing DNA chains of varying length that all terminate with the same 3′ base. These are separated by high-resolution polyacrylamide gel electrophoresis. In order to detect these chains, the most common methods employ the incorporation of a radiolabel or a fluorescent label. Radiolabels can be incorporated into the primer used for sequencing (by end-labeling with T4 polynucleotide kinase and [γ-^{32}P]ATP or [γ-^{33}P]ATP) or can be incorporated into the growing DNA chain by the use of, for example, [α-^{35}S]dATP or [α-^{32}P]dNTPs. [α-^{32}P]dNTPs have been used extensively for DNA sequencing; however, the high energy of the β particles that are emitted causes diffusion of the signals on autoradiography. These high-energy emissions also cause radiolysis of the DNA, which may cause problems when analyzing autoradiographs of sequencing gels. The introduction of [^{35}S]dATP [6] alleviated these problems because of the decreased scatter of the weaker β particles and the reduction in radiolytic damage. [^{35}S]dATP is used extensively for sequencing cloned PCR products, or for cycle-sequencing methods (see Section 7.4), but is not commonly used for the direct sequencing of PCR products. Recently, ^{33}P-labeled nucleotides ([γ-^{33}P]ATP, [α-^{33}P]dCTP, [α-^{33}P]dATP) have been reported for use in sequencing; ^{33}P has the advantage of being a stronger emitter than ^{35}S, resulting in a significant increase in sensitivity without loss of resolution. ^{33}P-labeled nucleotides are also weaker emitters than their ^{32}P-labeled equivalents and so handling is safer and the above problems are ameliorated by using this isotope.

As discussed above, radiolabeled oligonucleotide primers can be used for sequencing DNAs, including PCR products. [γ-^{32}P]ATP or [γ-^{33}P]ATP is available and the primer is labeled using polynucleotide kinase (see Section 2.4.3).

For fluorescent labels, dyes (e.g. JOE, ROX, FAM and TAMRA, P.E. Applied Biosystems) can be coupled to the primers which have been synthesized with a 5′ amino group (see Section 2.4.3). Each of these fluorescent dyes is color matched with one of the four dideoxy reactions, so each base termination is identified by a different fluorescence wavelength. Alternatively, the dyes can be attached to the ddNTPs (DyeDeoxy terminators™, P.E. Applied Biosystems) prepared using proprietary technology. These have the advantage of allowing the use of any primer for sequencing, avoiding the need to prepare the dye-labeled primers specifically.

Following autoradiography for radiolabeled sequencing reactions, or following fluorescent detection using an automated fluorescence-based DNA detection system, the DNA sequence can be determined. Fluorescence DNA detection systems can be either a DNA sequencer (e.g. ABI 373 or ABI Prism™ 377 DNA sequencers from P.E. Applied

Biosystems or ALF DNA Sequencer™ from Pharmacia) or a capillary electrophoresis system (ABI Prism™ 310 Genetic Analyzer). The 310 Genetic Analyzer is an automated system for sizing, sequencing and quantitation of nucleic acids. The system combines the ABI Prism™ multicolor fluorescent dyes and capillary electrophoresis using a novel separation medium (Genescan™ Polymer). The advantage of fluorescent DNA sequencing with DyeDeoxy terminators™ is that all the reactions can be performed in a single tube, since the DyeDeoxy terminators™ are labeled with different dyes that fluoresce at different wavelengths. Only fragments that have incorporated a dideoxynucleotide on their 3′ end carry a dye label, which can be detected individually using a real-time laser gel scanner. This reduces the amount of manual work involved and avoids track-to-track variation during electrophoresis. Radioactive sequencing and primer-labeled fluorescent sequencing requires the use of four separate sequencing reactions, one for each of the ddNTP terminators, and either four lanes or a single lane, respectively, on the polyacrylamide gel. Other methods, techniques and equipment for DNA sequencing are not reviewed here. Rather, the specific methods of sequencing PCR products are illustrated. Future advances in DNA sequencing technology should be applicable to sequencing PCR products (e.g. robotics for DNA template handling, the sequencing reactions and the use of separation techniques other than polyacrylamide gel electrophoresis).

7.1 Direct sequencing

Amplified DNA is generally double-stranded and linear. Alkaline denaturation is commonly used for circular plasmid DNA templates; however, this gives poor sequence with double-stranded PCR products. In order to denature the DNA strands, the PCR products are heated at 100°C for 5 min and then snap-cooled in a dry-ice/ethanol bath or liquid nitrogen, to prevent the strands from reannealing. Normally, a ^{32}P-labeled or fluorescently tagged (see Section 2.4.3) oligonucleotide primer is annealed to the denatured template and standard dideoxy sequencing methods are carried out using one of the DNA polymerases indicated above. Specific PCR products can be purified easily according to their size prior to sequencing, using, for example, agarose gel electrophoresis. If PCR products are suitably pure (i.e. a single band when checked by analytical agarose gel electrophoresis) they need only be freed of the low molecular weight primers and dNTPs. A number of rapid techniques can achieve this, for example centrifuge-driven spin dialysis (e.g. Centricon 100™ or Microcon 100™, Amicon Inc.), solid-phase purification if a suitably tagged PCR primer(s) was used (e.g. capture via biotin on magnetic beads coated with streptavidin), or polyethylene glycol/NaCl precipitation.

Direct DNA sequencing of PCR products permits the rapid characterization of sequences of interest without the need for subcloning [7–9]. There is also the advantage that PCR-derived errors in the amplification products are not detected as these should occur infrequently and at random positions in the amplified fragment. Sequencing of the large population of molecules present in the amplified DNA eliminates the chances of detecting PCR errors, unless the PCR was performed on very small amounts of template, in which case it is possible that a PCR error introduced early in the amplification could be detected in the final products. Therefore, direct sequencing gives the consensus sequence of the DNA molecules in the PCR reaction mixture. Examples of direct sequencing of PCR products with ^{32}P-labeled primers and fluorescently labeled DyeDeoxy terminators™ are shown in *Figures 7.1* and *7.2*, respectively.

7.2 Asymmetric PCR

In this approach, the addition of one primer in vast excess over the other during the PCR results in the generation of an excess of one amplified strand relative to the other, for example an asymmetric amplification. The single-stranded product can then serve as template for dideoxy sequencing [10,11]. An example of asymmetric PCR sequencing is shown in *Figure 7.1b*.

Asymmetric PCR can be performed directly on any given template, although reproducible high yields of single-stranded product are often difficult to obtain. Alternatively, it is more common to isolate a double-stranded PCR product by conventional symmetrical methods and reamplify this by asymmetric PCR, using only one primer for approximately 20 cycles. The single-stranded product increases arithmetically with each cycle; however, sufficient should be obtained for several sequencing reactions.

It is somewhat more difficult to determine the yield and quality of single-stranded product compared to double-stranded DNA. This quantification is usually done by either incorporation of a ^{32}P-labeled dNTP into the PCR product, Southern blotting the PCR product and probing with an oligonucleotide complementary to the single-stranded

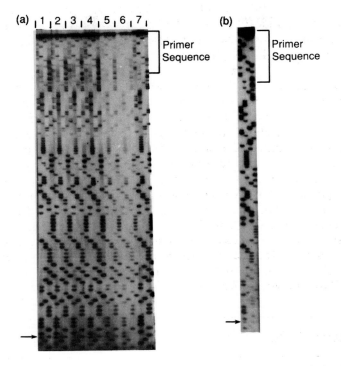

(a) 1 2 3 4 5 6 7

Primer Sequence

(b)

Primer Sequence

FIGURE 7.1: *Direct sequencing of double-stranded and single-stranded PCR products.* **(a)** *An autoradiograph of part of the DNA sequence of exon 5 of the alpha-1-antitrypsin gene from seven patients. These were obtained by direct sequencing of double-stranded PCR products using a* ^{32}*P-labeled oligonucleotide primer and T7 DNA polymerase (Sequenase®, USB). The arrow marks the position of the 'Z' mutation (GAG to AAG = Glu^{342} to Lys^{342}) which is a common cause of alpha-1-antitrypsin deficiency. Examples of the three possible genotypes are shown. GAG is normal (DNAs 3 and 7), AAG is homozygous Z (DNA 6) and G and A/AG are heterozygous (DNAs 1, 2, 4 and 5). The position of the primer at the end of the PCR product is shown and for each sequencing track the lanes are, left to right, A, C, G and T.* **(b)** *DNA sequencing of exon 5 of the alpha-1-antitrypsin gene obtained by direct sequencing of a single-stranded asymmetric PCR product using a* ^{32}*P-labeled oligonucleotide primer. The arrow marks the position of the 'Z' mutation, this patient is heterozygous at this locus. The position of the primer at the end of the PCR product is shown. The lanes are, left to right, A, C, G and T.*

product, or by ethidium bromide staining of products electrophoresed in agarose or polyacrylamide gels.

Dideoxy DNA sequencing of asymmetric PCR products can be performed with either the original PCR primer (but not the one used

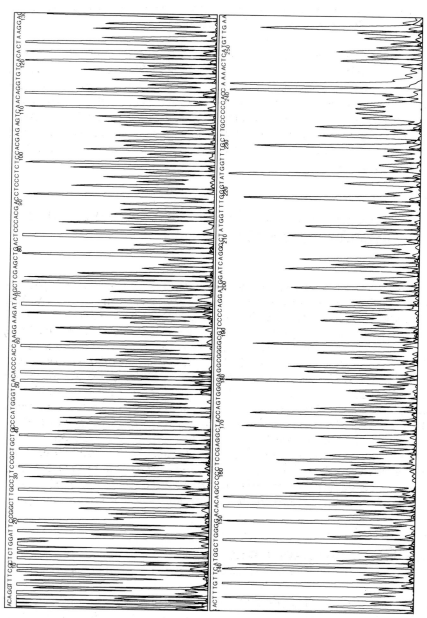

FIGURE 7.2: *Processed data from automated DNA sequence analysis of a PCR product (human neurokinin A receptor cDNA). A 1278 bp PCR product was purified by spin dialysis (Microcon 100™ Amicon Inc.) and cycle sequenced using an internal primer and Amplitaq® Taq cycle sequencing kit (P.E. Applied Biosystems). Fluorescence-based DNA sequence analysis was performed using a P.E. Applied Biosystems 373 DNA sequencer with a 6% polyacrylamide gel, using the manufacturer's version 1.2.0. software.*

for the asymmetric PCR) or by using a complementary sequence internal to the asymmetric PCR product.

Asymmetric PCR sequencing has the advantage of being able to use ^{35}S-labeled dNTPs in place of ^{32}P-labeled dNTPs, and also gives a sequence of excellent quality. This method is particularly useful for repetitive sequencing of a particular region where the asymmetric PCR conditions can be standardized, whereas generating single-stranded PCR products from different target sequences can often require optimization for each, particularly with respect to the ratios of the two primers.

7.3 Single-stranded DNA sequencing using lambda exonuclease III

Single-stranded products can be generated by the selective digestion of one of the two strands of a double-stranded amplified product [12]. Lambda exonuclease III is a double-strand-specific 5′ to 3′ exonuclease, which requires the presence of a phosphate group at the 5′ position for activity. By using one PCR primer which is 5′ phosphorylated, either chemically during its synthesis or post-synthesis using T4 polynucleotide kinase and ATP, the 5′ end of the PCR product strand extended from this primer will be phosphorylated. This strand will be degraded when treated with lambda exonuclease III. DNA sequencing may then be performed as described for asymmetric PCR products (see Section 7.2) and shares the same technical benefits as asymmetric PCR sequencing. The single-stranded templates can be sequenced by incorporating any labeled nucleotide in the labeling reaction and the use of single-stranded DNA eliminates the denaturation and precipitation steps.

7.4 Cycle sequencing

Cycle sequencing (*Figure 7.3*) involves the linear amplification of double-stranded DNAs (e.g. PCR products, plasmids, cosmids) or single-stranded products (e.g. asymmetric PCR products). For symmetric PCR products either strand can be sequenced and this requires no titration of primer concentrations (as is required in asymmetric PCR). The thermocycler performs the denaturation, annealing and extension steps of a typical PCR, but because only a

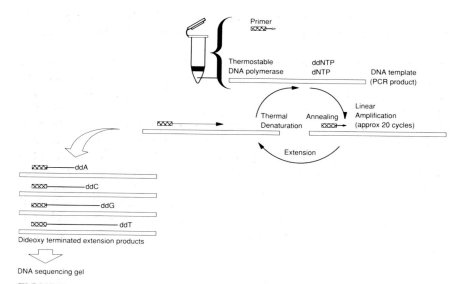

FIGURE 7.3: *Cycle sequencing. PCR products (double-stranded or asymmetric) are mixed with a single primer, dNTPs, ddNTPs and a thermostable DNA polymerase. Linear amplification is performed in a thermal cycler for approximately 20 cycles. Each cycle consists of annealing, extension and thermal denaturation as in a normal PCR. Protocols are available for the use of radiolabeled or fluorescent dye-labeled primers. Alternatively, radiolabeled deoxynucleotides or fluorescent dye-labeled ddNTPs may be used. The sequencing reactions can be analyzed by polyacrylamide/urea gels followed by detection with X-ray film or automated fluorescence sequencers, as appropriate.*

single primer is present in the reaction this results in a linear amplification. The extension products are then applied to a sequencing gel. Radioactive or automated fluorescent sequencing can be performed. Cycle sequencing employs a single procedure for all templates, and results in a longer, more accurate length of readable sequence than alternative PCR sequencing methods. It also allows the sequencing of unpurified DNA, such as plasmid from bacterial cell lysates. This is due to the fact that the process results in amplification, albeit linear, therefore the substrate can be significantly diluted to reduce the concentrations of possible inhibitors of the sequencing reaction.

Several thermostable DNA polymerases have been reported to be useful for cycle sequencing and are commercially available as part of sequencing kits (e.g. Amplitaq®, P.E. Applied Biosystems; VentR™(exo⁻), New England Biolabs; Cyclist exo⁻ *Pfu*, Stratagene).

Recently, a mutant form of Amplitaq (Amplitaq® DNA polymerase, FS (Fluorescent Sequencing)) has been developed by Roche Molecular Systems, and is commercially available from P.E. Applied Biosystems, for improved sequencing of a variety of single- and double-stranded DNA templates. This genetically modified enzyme has very low 5' to 3' nuclease activity and this results in reduced background noise and very few false terminations. In addition, Amplitaq® FS incorporates ddNTPs much more efficiently than other enzymes and so lower concentrations of ddNTPs can be used in both dye primer and dye terminator reactions. This improves overall enzyme efficiency and performance and, as a result of the greater fluorescent signal, less starting DNA is required and dye removal before gel loading is simplified.

7.5 Direct solid-phase sequencing

PCR products can be sequenced by capture of one of the amplicon strands to a solid phase, followed by denaturation and DNA sequencing of the immobilized strand. One of the primers used for PCR is biotinylated at the 5' end (see Section 2.4.3) and following PCR amplification the product is captured with magnetic beads covalently coupled to streptavidin [13–15] or with a streptavidin affinity gel [16,17]. The nonbiotinylated strand is eluted with alkali to yield single-stranded DNA immobilized at the 5' end. A sequencing primer can then be used for solid phase dideoxy sequencing. Either a ³²P or fluorescently end-labeled primer can be used; alternatively, radiolabeled dNTPs or fluorescently labeled ddNTPs can be used in the sequencing reactions. The scheme of this sequencing method is shown in *Figure 7.4*.

Solid-phase sequencing has many advantages compared to alternative PCR sequencing strategies. Sequencing from both ends of the amplicon is easily achieved as either of the two primers used in the PCR can be separately biotinylated. The magnetic or affinity support facilitates handling and purification and allows optimal buffer conditions for both amplification and sequencing by simple buffer exchange. Solid-phase sequencing has the potential to be automated and to provide a simple method for large sequencing projects.

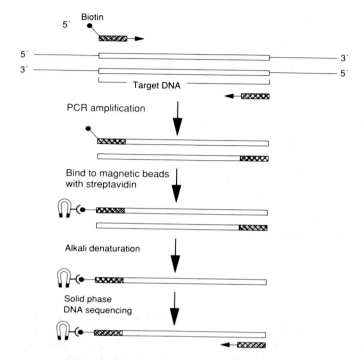

FIGURE 7.4: *Direct solid-phase sequencing of PCR products. PCR products are captured after amplification via a biotin label to streptavidin coupled to magnetic beads. Following alkali denaturation, the immobilized DNA is sequenced.*

7.6 Gene amplification with transcript sequencing (GAWTS)

This sequencing technique is based on the amplification of genomic sequence by PCR, followed by transcription and sequencing of the transcripts (see *Figure 7.5*) [18]. A bacteriophage promoter, for example the T7 promoter, is attached to the 5′ end of at least one of the PCR primers, as discussed in Section 4.2.1. After amplification, the PCR products are transcribed into multiple copies of RNA using T7 RNA polymerase. This increases the signal and provides single-stranded template for dideoxy sequencing with reverse transcriptase. This is achieved using an internal sequencing primer to prime the reverse transcription. GAWTS is a rapid technique as it is not necessary to clone PCR products to add the promoter sequence; it is

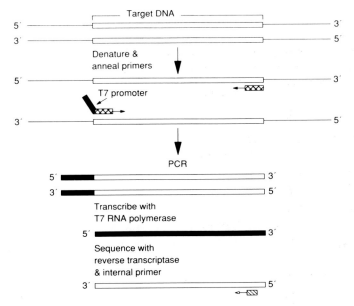

FIGURE 7.5: *Genomic amplification with transcript sequencing (GAWTS).*
Genomic amplification is performed where one of the primers contains a
bacteriophage promoter sequence (T7 shown here) at its 5′ end. Following
PCR, the product can be transcribed using T7 RNA polymerase and the
RNA then dideoxy sequenced using reverse transcriptase.

also very sensitive, as little as 1 ng of genomic DNA can give sequence information.

More than one region of genomic DNA can be simultaneously amplified and transcribed and then each sequence determined by using appropriate reverse transcriptase primers. This technique is suitable for automation as ethanol precipitation and centrifugation steps are not required; the sensitivity is high due to the secondary amplification through transcription from the bacteriophage RNA polymerase. The quality of the sequence information is also high due to sequencing of a single-stranded template.

A modification of GAWTS has also been reported [19], RNA amplification with transcript sequencing (RAWTS), whereby cDNA is synthesized from RNA with oligo dT, random hexamers or an mRNA-specific oligonucleotide primer prior to GAWTS. This technique has shown that basal levels of tissue-specific mRNA can be detected. Potentially this allows the study of mRNA in patients with genetic diseases, rather than their genomic DNA which can be problematic when a large complex gene is being studied.

It is also possible to attach a translation initiation signal to the 3′ end of the bacteriophage promoter sequence of the PCR primer. If the PCR product is then transcribed in the presence of a capping agent such as 7mGpppG, a capped RNA is produced, which can be translated into protein by *in vitro* translation using wheat-germ or rabbit reticulocyte lysate. This technique, known as RNA amplification with *in vitro* translation (RAWIT) [19], is a powerful method for examination of structure–function relationships of proteins, where the function of the protein can be measured after *in vitro* translation.

7.7 Chemical sequencing

In chemical sequencing, otherwise known as the Maxam–Gilbert technique [2], single end-labeled DNA is partially degraded using a set of base-specific cleavage reactions. The reaction conditions are chosen in such a way that only a limited number of cleavages occur in each DNA molecule. The four bases are modified in separate reactions: G is methylated using dimethylsulfate; G and A are methylated using formic acid; C and T are removed with hydrazine; C alone is removed by hydrazine in the presence of NaCl. Other base-specific reactions have also been reported in the literature. The separate reactions are then treated with piperidine, which cleaves the DNA in the sugar–phosphate backbone where there is a missing or modified base. The radiolabeled cleaved fragments that are produced are analyzed on standard DNA sequencing gels.

For chemical sequencing it is necessary to end-label one of the strands of double-stranded DNA. For PCR sequencing this can easily be done by 5′ end-labeling one of the PCR primers. This results in amplification of double-stranded product containing one labeled and one unlabeled strand. The labeled PCR product is purified by agarose or polyacrylamide gel electrophoresis to remove unincorporated labeled primer and [γ-32P]ATP from the polynucleotide kinase reaction (see Section 2.4.3). The PCR product is then isolated from the gel using one of a number of standard methods for fragment purification.

Chemical sequencing of PCR products [20–22] is less commonly used than dideoxy sequencing as it involves more steps, requires radiolabeled primers and necessitates the use of hazardous chemicals. The chemicals used for this method of sequencing are extremely toxic and should be used exclusively within a fume cupboard. For these reasons chemical sequencing is rarely the method of choice given the other safer and simpler methods now available.

7.8 PCR archaeology

DNA has been extracted and amplified from several ancient biological remains (*Table 7.1*). However, there are only two reports of molecular cloning of such remains [23, 24] prior to PCR. Since the advent of PCR many such archaeological remains have been amplified and sequenced. Many of these studies have focused on the amplification of mitochondrial (mt) or chloroplast genes which are present at hundreds to thousands of copies per cell, compared with one or two per cell for genes found in the nucleus. This is important when one considers the extent of damage to DNA caused by the death of the organism and the passage of time. Therefore the survival of some intact DNA from these organelles is more likely than that of nuclear DNA. Mitochondrial DNA is also informative from a phylogenetic viewpoint [25], as an individual usually contains only one type of molecule. Obviously the study of mtDNA has limitations since it only encodes a relatively small number of genes and these are non-nuclear and inherited maternally. However, several groups have recently reported the amplification of nonmitochondrial genes from ancient remains and museum specimens.

TABLE 7.1: *Examples of archaeological specimens used for PCR*

Sample origin	Gene amplified	Size of amplicon (bp)	Age of sample (years)	Reference
Marsupial wolf	mtDNA (cyt *b*) mtDNA (12S rRNA)	116 94	~100	26
Human cortical bone	mtDNA mtDNA (NADH dehydrogenase)	121 205	300–5500	27
Human brain	mtDNA (cyt b_2)	471	7000	28
Human brain	HLA and β_2-microglobulin	several <125	7500	29
Alaskan horse bone	(16S rRNA)	91	25000	30
Siberian woolly mammoth	mtDNA (cyt *b*)	375	47000	31, 32
Magnolia fossil leaf	chloroplast gene (rbc L)	820	17–20×10^6	33
Stingless bee in amber	nuclear (18S rRNA)	555, 195 and 597	30×10^6	34
Termite in amber	mtDNA (16S rRNA) nuclear (18S rRNA)	150 225	30×10^6	35
H. protera leaf in amber	chloroplast gene (rbc L)	346	35–40×10^6	36
Weevil in amber	18S rRNA	315 and 226	120–135×10^6	37

The main problem in this field is in proving the authenticity of the PCR products and their DNA sequences and some of the pitfalls are discussed [38, 39]. Because of the great sensitivity of PCR, it is possible to envisage contamination caused by the handling of ancient specimens, bacterial or fungal contamination, cross-contamination from other species or with other biological samples handled in the laboratory. It is crucial that great care is taken in designing experiments to look at archaeological remains. Ideally, the sequences obtained would be similar but not identical to living species. Independent confirmation of the same sequences by another laboratory, or the use of Y-chromosome-specific sequences to confirm the sex of human skeletal remains of known sex are examples of the sort of controls that will help in this type of study. Conventional controls such as amplification without extract and multiple extracts from the same specimen are also crucial. Despite these difficulties, the field of PCR archaeology is on the increase and it opens up the possibility of using museum specimens, as well as archaeological finds, to address questions of evolutionary, taxonomic and historic significance.

References

1. Sanger, F., Nicklen, S. and Coulson, A.R. (1977) *Proc. Natl Acad. Sci. USA,* **74,** 5463.
2. Maxam, A.M. and Gilbert, W. (1977) *Proc. Natl Acad. Sci. USA,* **74,** 560.
3. Tabor, S. and Richardson, C.C. (1987) *Proc. Natl Acad. Sci. USA,* **84,** 4767.
4. Innis, M.A., Myambo, K.B., Gelfand, D.H. and Brow, M.A.D. (1988) *Proc. Natl Acad. Sci. USA,* **85,** 9436.
5. Mead, D.A., McClary, J.A., Luckey, J.A., Kostichka, A.J., Witney, F.R. and Smith, L.M. (1991) *BioTechniques,* **11,** 76.
6. Biggin, M.D., Gibson, T.J. and Hong, G.F. (1983) *Proc. Natl Acad. Sci. USA,* **80,** 3963.
7. McMahon, G., Davis, E. and Wogan, G.N. (1987) *Proc. Natl Acad. Sci. USA,* **84,** 4974.
8. Wong, C., Dowling, C.E., Saiki, R.K., Higuchi, R.G., Erlich, H.A. and Kazazian, H.H. (1987) *Nature,* **330,** 384.
9. Wrischnik, L.A., Higuchi, R.G., Stoneking, M., Erlich, H.A., Arnheim, N. and Wilson, A.C. (1987) *Nucleic Acids Res.,* **15,** 529.
10. Gyllensten, U.B. and Erlich, H.A. (1988) *Proc. Natl Acad. Sci. USA,* **85,** 7652.
11. Gyllensten, U.B. (1989) in *PCR Technology* (H.A. Erlich, ed.). Stockton Press, New York, pp. 45–60.
12. Higuchi, R.G. and Ochman, H. (1989) *Nucleic Acids Res.,* **17,** 5865.
13. Hultman, T., Stahl, S., Hornes, E. and Uhlen, M. (1989) *Nucleic Acids Res.,* **17,** 4937.
14. Hultman, T., Bergh, S., Moks, T. and Uhlen, M. (1991) *BioTechniques,* **10,** 84.

15. Kaneoka, H., Lee, D.R., Hsu, K.-C., Sharp, G.C. and Hoffman, R.W. (1991) *BioTechniques,* **10,** 30.

16. Stahl, S., Hultman, T., Olsson, A., Moks, T. and Uhlen, M. (1988) *Nucleic Acids Res.,* **16,** 3025.

17. Mitchell, L.G. and Merril, C.R. (1989) *Anal. Biochem.,* **178,** 239.

18. Stoflet, E.S., Koeberl, D.D., Sarkar, G. and Sommer, S.S. (1988) *Science,* **239,** 491.

19. Sarkar, G. and Sommer, S.S. (1989) *Science,* **244,** 331.

20. DiMarzo, R., Dowling, C.E., Wong, C., Maggio, A. and Kazazian, H.H. (1988) *Br. J. Haematol.,* **69,** 393.

21. Ohara, O., Dorit, R.L. and Gilbert, W. (1989) *Proc. Natl Acad. Sci. USA,* **86,** 5673.

22. Kraus, J.P. and Tahara, T. (1993) *Methods Enzymol.,* **218,** 227.

23. Higuchi, R., Bowman, B., Freiberger, M., Ryder, O.A. and Wilson, A.C. (1984) *Nature,* **312,** 282.

24. Pääbo, S. (1985) *Nature,* **314,** 644.

25. Wilson, A.C., Cann, R.L., Carr, S.M., George, M., Gyllensten, U.B., Helm-Bychowski, K.M., Higuchi, R.G., Palumbi, S.R., Prager, E.M., Sage, R.D. and Stoneking, M. (1985) *Biol. J. Linn. Soc.,* **26,** 375.

26. Thomas, R.H., Schaffner, W. and Wilson, A.C. (1989) *Nature,* **340,** 465.

27. Hagelberg, E., Sykes, B. and Hedges, R. (1989) *Nature,* **342,** 485.

28. Pääbo, S., Gifford, J.A. and Wilson, A.C. (1988) *Nucleic Acids Res.,* **16,** 9775.

29. Lawlor, D.A., Dickel, C.D., Hauswirth, W.W. and Parham, P. (1991) *Nature,* **349,** 785.

30. Höss, M. and Pääbo, S. (1993) *Nucleic Acids Res.,* **21,** 3913.

31. Höss, M., Pääbo, S. and Vereshchagin, N.K. (1994) *Nature,* **370,** 333.

32. Hagelberg, E., Thomas, M.G., Cook, C.E.J., Jr, Sher, A.V., Baryshnikov, G.F. and Lister, A.M. (1994) *Nature,* **370,** 333.

33. Golenberg, E.M., Giannasi, D.E., Clegg, M.T., Smiley, C.J., Durbin, M., Henderson, D. and Zurawski, G. (1990) *Nature,* **344,** 656.

34. Cano, R.J., Poinar, H.N. and Poinar, G.O., Jr (1992) *Med. Sci. Res.,* **20,** 619.

35. De Salle, R., Gatesy, J., Wheeler, W. and Grimaldi, D. (1992) *Science,* **257,** 1933.

36. Poinar, H.N., Cano, R.J. and Poinar, G.O., Jr (1993) *Nature,* **363,** 677.

37. Cano, R.J., Poinar, H.N., Pieniazek, N.J., Acra, A. and Poinar, G.O., Jr (1993) *Nature,* **363,** 536.

38. Lindahl, T. (1993) *Nature,* **365,** 700.

39. Poinar, G.O., Jr (1993) *Nature,* **365,** 700.

8 New Sequence Determination and Genome Mapping

It is frequently necessary to determine the DNA sequences of regions adjacent to known DNA sequences. The reasons include sequence elucidation *per se*, the generation of end-specific probes for cosmids, yeast artificial chromosomes (YACs), P1-derived artificial chromosomes (PACs) and bacterial artificial chromosomes (BACs) and the study of DNA sequences that can integrate or transpose into the genome. This chapter describes the methods available for characterizing regions of unknown DNA sequence that are adjacent to stretches of previously characterized DNA sequence, generating chromosomal and genomic libraries, physical mapping, chromosome walking and the generation of sequence tagged sites (STSs).

8.1 New sequence by vectorette PCR

Vectorette PCR, otherwise known as 'chemical genetics' or 'bubble PCR' [1, 2] is a technique that utilizes a ligated, partially mismatched oligonucleotide 'vectorette' as the priming site for one of the PCR primers. The other priming site is provided by the stretch of known DNA sequence. Because the vectorette unit comprises a partially mismatched oligonucleotide duplex (hence the term 'bubble PCR'), its PCR primer cannot anneal and extend until the fully complementary sequence is provided by extension of the primer specific for the known sequence (*Figure 8.1*). Thus, restriction-enzyme-digested DNA ligated to a vectorette unit can be regarded as a library from which any DNA sequence can be characterized. The only provision is that the ligated vectorette unit is within a PCR-amplifiable distance from the segment of known sequence. In practice, several vectorette libraries are usually prepared using different restriction enzymes. This will

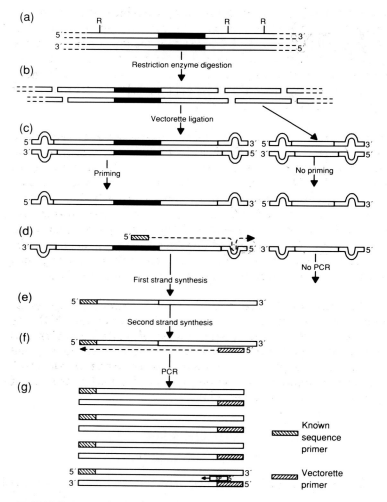

FIGURE 8.1: *(a)* *Duplex DNA with a region of known nucleotide sequence and flanked by restriction enzyme recognition sites, R.* *(b)* *Restriction digestion cleaves the duplex DNA into fragments.* *(c)* *The vectorette unit is ligated to all DNA fragments from the restriction digest.* *(d)* *A primer complementary to one strand of the known sequence is extended during the first-strand synthesis during the first PCR cycle, generating a product whose 3' end is now fully complementary to the vectorette ligated to the template strand. Where the known-sequence primer does not bind, there is no complement to the ligated vectorette produced.* *(e)* *and* *(f)* *Second-strand synthesis in the second PCR cycle can only occur where the vectorette complementary sequences have been produced (i.e. on extension products of the known-sequence primer).* *(g)* *Subsequent cycles of PCR amplify DNA bounded by the known-sequence primer and the vectorette primer; direct sequencing can be performed using another primer (SP) designed from the original vectorette unit.*

maximize the probability of generating an appropriate-sized fragment of DNA flanked by a restriction enzyme recognition site and the known region of DNA sequence. PCR products derived from vectorette PCR may be sequenced as described in Chapter 7, using a primer complementary to an internal portion of the vectorette. The design of a typical vectorette unit and its associated PCR primers and sequencing primer is shown in *Figure 8.2*.

A modification of the vectorette PCR, double-ended vectorette incorporating alternative transcription sites (DEVIATS), requires the ligation of two different vectorette units to allow the sequencing of both DNA strands adjacent to the initial vectorette unit. In this method [3] a vectorette library is prepared as described and, after restriction digestion with a different enzyme whose recognition site occurs between the known sequence and the first ligated vectorette unit, a second vectorette is added. After PCR using primers specific for the two vectorette units, sequencing may be performed using both vectorette sequencing primers in separate sequencing reactions.

As with PCR in general, the complexity of whole genomes may result in the generation of nonspecific products in vectorette PCR. As discussed in Section 3.7, the implementation of nested primers will usually obviate this problem. Therefore, this should be considered during the design of the primer(s) complementary to the known DNA sequence. Nested primers may also be used for the vectorette end of the PCR product, as shown in *Figure 8.2*.

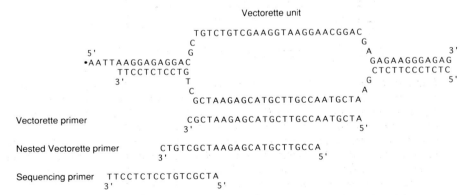

FIGURE 8.2: *An* EcoRI *vectorette unit;* • *indicates a 5′ phosphate group on the* EcoRI *cohesive end. The vectorette has a mismatched central (bubble) region. The vectorette, nested vectorette and sequencing primers are all complementary to a portion of the sequence that would be generated after the bottom vectorette strand had been incorporated into the PCR product during first-strand synthesis.*

Naturally, after having determined the DNA sequence adjacent to the known sequence starting point, it is possible to design additional specific primer sets. Vectorette libraries may then be amplified again with these primers and the original vectorette PCR primers. In this way, it is possible to continue to 'walk' along the genome from the original limited sequence information.

8.2 New sequence by inverse PCR

Inverse PCR [4] is a system that amplifies unknown DNA sequence either side of a core region of known sequence. The principle of this method is the circularization of linear restriction enzyme fragments that contain the known DNA sequence. Amplification of such circularized DNAs with primers designed specifically for the known sequence and oriented outwards with respect to the known region, gives a product of the unknown sequences that flank the region of known sequence. This amplified unknown sequence comprises the two flanking sequences joined at the restriction enzyme site regenerated during the circularization step (*Figure 8.3*). Sequencing may then be carried out from either end of the PCR product using primers designed from the known sequence. When the regenerated restriction enzyme recognition site is reached, the external boundary of contiguous new sequence is defined for that sequencing primer. Any new sequence read past the restriction enzyme recognition site is derived from the extreme end of the original restriction fragment. It is therefore to the other side of the originally known sequence and in the inverted orientation.

8.3 PCR screening and mapping of cosmids, yeast artificial chromosomes (YACs), P1-derived artificial chromosomes (PACs) and bacterial artificial chromosomes (BACs)

Cosmid cloning systems are based on modified plasmids that are capable of accepting relatively large fragments of DNA up to 50 kb. Cosmids contain one or more *cos* sites from the bacteriophage λ. They also carry an origin of replication and a selectable antibiotic

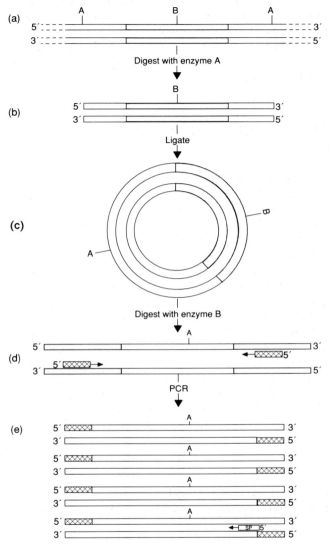

FIGURE 8.3: **(a)** *Duplex DNA with a region of known nucleotide sequence containing restriction enzyme recognition site B and flanked by restriction enzyme recognition sites, A, within regions of unknown sequence.*
(b) *Restriction digestion with enzyme A cleaves the duplex DNA into fragments.* **(c)** *Ligation of the linear fragments to generate circular DNA molecules.* **(d)** *Restriction digestion with enzyme B opens the circular molecules to recreate linear duplexes where the unknown sequence is now flanked by portions of the known sequence.* **(e)** *PCR using primers designed from the known sequence generates a PCR product containing two regions of unknown sequence converged at restriction recognition site A. SP=sequencing primer.*

resistance marker for propagation as a plasmid. The presence of the *cos* site(s) allows packaging of individually cloned DNA fragments into the head of bacteriophage λ, consequently, genomic DNA libraries constructed in cosmids can be preserved as stocks of phage particles. YAC libraries are constructed using plasmid vectors that carry yeast chromosome telomere sequences and a yeast selectable marker [5]. In cloning genomic DNA fragments, that may be up to 1 Mb, the vector design gives rise to the telomeric sequences at either end of a linear DNA molecule that can be maintained in yeast by virtue of the selectable marker. The bacteriophage P1 can be used as a cloning system that is capable of receiving inserts of genomic DNA up to 100 kb in size [6] and as such complements cosmids and YACs which are capable of accepting inserts up to 50 kb and 1 Mb, respectively. The vector used in constructing PAC clones and libraries comprises the bacteriophage *pac* site on one vector arm which initiates packaging of ligation products. The other vector arm contains the genetic information for selection and replication of the P1 plasmid clone. T7 and SP6 promoters are also present, allowing the generation of RNA probes for chromosome walking and DNA sequencing. A bacterial system (BAC) for cloning and mapping DNA fragments in the region of 300 kb has been derived from *E. coli* and its single-copy plasmid F factor [7]. The BAC system plasmid carries the regulatory genes from the plasmid F factor, thus maintaining single-copy status of individual cloned DNA fragments so providing stability to cloned DNAs. The plasmid is maintained by the presence of an antibiotic-resistance marker. T7 and SP6 promoters are also present for the same reasons as described for PAC clones above. Each of the above systems has been applied to creating chromosome-specific libraries and whole genome libraries.

The isolation of terminal sequences and STSs from genomic clones is an essential aspect of genome walking and contig formation, for example the progressive characterization of genomic DNA by isolating adjacent and overlapping nucleotide sequences of cloned DNA. STSs are sequences that can be specifically amplified by PCR. Genomic clones sharing one or more STSs can therefore be ordered to generate contigs quickly and without recourse to DNA sequencing for contig alignment. This STS-based contig assembly approach was applied to the generation of a 2.6 Mb contig spanning the dystrophin gene; the assembly of this contig required just 27 STSs [8]. Contigs can be screened subsequently for expressed sequences, gene exons, which in turn can be used to identify full-length cDNA clones and then mutations responsible for genetic diseases. Another routine requirement is the identification of chimeric clones, a common artifact of many libraries. Chimeric clones need to be identified since they contain noncontiguous pieces of DNA and therefore create problems in genome walking and physical mapping projects. The isolation of terminal sequences using the vectorette or inverse PCR methods, the generation of PCR products for probes based on these sequences, and hybridization to somatic cell monochromosome hybrid cell libraries allow the detection of these chimeric clones [9]. An alternative method involves Alu PCR (see Section 8.5), where Alu primers are used in conjunction with primers for the arm sequences of the vector; however, this relies on there being Alu sequences within PCR range of the vector primers. In some instances Alu PCR may be the only method of screening YAC ends if Alu repetitive sequences separate the YAC arm from restriction sites suitable for other methods; this is, however, a rare occurrence.

8.4 Chromosomal localization

The conventional method for assignment of a specific DNA sequence to a particular human chromosome is based on Southern blot hybridization analysis of genomic DNAs from human–rodent somatic cell hybrids. Each somatic cell hybrid contains the normal complement of rodent chromosomes along with one or more human chromosomes which can be identified by a number of techniques. This method requires large amounts of DNA (5–10 µg from >30 hybrid cell lines), is time consuming and requires hybridization with radio-labeled probes. In a PCR-based approach [10], PCR is used to specifically amplify human sequences and not their rodent homologs. For this reason it is advisable to select primers that are selective for the human gene (e.g. from intron sequences or untranslated ends of genes), particularly if the gene of interest belongs to a highly conserved family of genes. PCR of rodent DNA versus human genomic DNA is an important control for checking the specificity of the primers and for optimizing the PCR conditions. An example of chromosome assignment is shown in *Figure 8.4*.

Many research laboratories have prepared somatic cell hybrid panels; however, these should be carefully validated experimentally as being suitable for analysis involving PCR. A human–hamster somatic panel of PCRable™ DNAs is available from Bios Corporation. Mapping panels are also available from CORIELL Repository, Camden, New Jersey, USA. Alternatively, several laboratories have prepared monochromosome panels (e.g. human–mouse) [11], and potentially these could be used for PCR-based chromosome assignment. The PCR technique can be used to map the position of a gene on a particular chromosome more precisely, by using panels of somatic cell hybrids

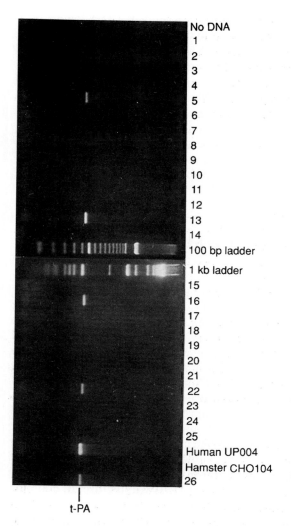

No DNA
1
2
3
4
5
6
7
8
9
10
11
12
13
14
100 bp ladder
1 kb ladder
15
16
17
18
19
20
21
22
23
24
25
Human UP004
Hamster CHO104
26

t-PA

FIGURE 8.4: *Chromosome mapping by PCR. PCR of human tissue-type plasminogen activator (t-PA) gene on somatic cell hybrids and their parent cell lines. Primers specific for the 3' end of the human t-PA gene (in exon XIV) were used for PCR and the 537 bp products were analyzed by agarose gel electrophoresis. PCRs 1–25 correspond to hamster–human hybrids (Bios Corporation) and PCR 26 corresponds to a mouse–human monochromosome hybrid known to contain human chromosome 8 (E. Stanbridge, University of California, Irvine). Control PCRs with human DNA (UP004), hamster DNA (CHO104) and no DNA are also shown. Size markers are 1 kb ladder and 100 bp ladder (Gibco-BRL).*

which have a set of well-characterized, selectively retained, translocated portions of any single chromosome. Chromosome assignment can also be used as a method of validation during chromosome-walking techniques. Often the ends of a cosmid or YAC clone have been determined and the new fragments or sequences will be used as probes for walking, and it is useful to confirm that all the new sequences are indeed from the expected chromosome.

8.5 Alu PCR

Alu PCR utilizes the ubiquitous Alu repeat sequence found in human DNA. There are in the region of 10^6 copies of this c. 300 bp sequence dispersed throughout the genome [12]. There is variation between the copies of this sequence, but a consensus sequence has been established and there are regions of the repeat that are very well conserved and from which PCR primers can be designed [13].

Alu primers, combined with vector-specific primers adjacent to cloning sites, allow the direct isolation of human inserts in cloned DNAs (see Section 8.3). These can, in turn, be used as probes for chromosomal localization studies (see Section 8.4). Alu primers may also be used independently of vector-specific primers for the amplification of inter-Alu sequences from either cloned or uncloned genomic DNA.

Band patterns produced by Alu PCR can also prove useful in the analysis of human regions in somatic cell hybrids that retain human chromosome fragments in a rodent cell background. This is because in the second Alu repeat monomer there is a region of 31 nucleotides that is unique to primates. Alu PCR can therefore be used to 'fingerprint' these, allowing the overlaps in hybrids or clones to be determined. This difference between human and rodent Alu repeats means that Alu sequences can also be used specifically to amplify human DNA from the somatic cell hybrids. Therefore, specific human regions may be characterized without the need for genomic DNA libraries, although the technique can also be used with these libraries. Alu PCR using vector-specific primers, as discussed above, can also be used in conjunction with hybridization of the PCR product to somatic cell hybrids for chromosomal localization of cloned genomic fragments. This therefore saves considerable time as independent clones do not have to be grown and DNA isolated from them to prepare hybridization probes.

8.6 Transgenic animal screening

The construction of transgenic animals by the direct microinjection of DNA fragments into embryos is an established technique for functional and developmental analysis. Rapid screening of transgenes is important to allow rapid breeding of transgenic animal colonies and to avoid the high costs of keeping untested animals. Therefore, the use of PCR to replace Southern blot analysis has simplified and increased the throughput, especially where several litters have to be analyzed. Generally, the screening of transgenic animals, particularly the back-crossing of transgenes from a founder animal into an inbred background is laborious. However, PCR has simplified the detection of specific DNA sequences in this context.

In mice, only 0.1% of the blood cells are nucleated leukocytes. The remaining 99.9% are non-nucleated erythrocytes from which DNA cannot be isolated. Given that it is difficult to obtain significant quantities of blood from mice, the sensitivity of PCR makes it the ideal tool for screening. Not only can transgenes be detected in adult mice (which have around 10^4 leukocytes μl^{-1} of blood) but they can be detected in fetuses which have only one-tenth of the number of leukocytes per μl of blood compared to the adult animal. Similarly, transgenes can be detected from mice used in bone marrow transplant experiments where low cell counts are a problem. Typically, several litters of mice can be screened per day using a single drop of blood as the source of the DNA template, making the process more humane than the Southern blot screening procedures, which require sections of tail in order to isolate sufficient DNA to perform an analysis. Even more humane and simple to perform is the recently reported use of mouse saliva [14] to screen for transgene integration.

8.7 PCR amplification of microdissected DNA

Several groups have reported molecular cloning of specific chromosomal regions from humans and the fruit fly *Drosophila melanogaster* by microdissection of chromosomes and amplification by

PCR [15–17]. This technique has been termed chromosome microdissection PCR (CM-PCR), and it is particularly useful for preparing libraries of specific chromosomal regions and in gene-mapping studies. Different methods have been used for preparing libraries by CM-PCR that allow the universal amplification of the microdissected DNA [15,16]. More than 80% of the clones obtained were single copy and could be localized to the dissected region by *in situ* hybridization. This technique is a powerful primary step in gene identification and physical mapping. A collection of PCR-amplified fragments from a specific chromosome or region of a chromosome can be labeled either isotopically or nonisotopically and used to 'paint' the chromosomes by *in situ* hybridization for karyotype analysis. Structural genes and other characterized DNA sequences can be mapped on to chromosomes by several methods, including *in situ* hybridization, the use of somatic cell hybrids (see Section 8.4) and fluorescence-activated chromosome sorting. CM-PCR has also been used to map genes [15] and the technique has several advantages over traditional methods of gene mapping. The process is rapid and a PCR result should be obtained within a day of the microdissection. Potentially, the level of resolution is as good as that with isotopically labeled probes used by *in situ* hybridization, and the technique may be used with archival metaphase chromosome spreads. CM-PCR may also benefit other types of genetic studies and cancer cytogenetics. Furthermore, this technique provides the researcher or clinician with the ability to study DNA sequences from a single chromosome and to distinguish homologous chromosomes.

References

1. Riley, J., Butler, R., Finniear, R., Jenner, D., Powell, S., Anand, R., Smith, J.C. and Markham, A.F. (1990) *Nucleic Acids Res., 18*, 2887.
2. Arnold, C. and Hodgson, I.J. (1991) *PCR Methods Appl., 1*, 39.
3. Copley, C.G., Boot, C., Bundell, K. and McPheat, W.L. (1991) *BioTechnology, 9*, 74.
4. Triglia, T., Peterson, M.G. and Kemp, D.J. (1988) *Nucleic Acids Res., 16*, 8186.
5. Burke, D.T., Carle, G.F. and Olsen, M.V. (1987) *Science, 236*, 806.
6. Sternberg, N. (1990) *Proc. Natl Acad. Sci. USA, 87*, 103.
7. Shizuya, H., Birren, B., Kim, U.-J., Mancino, V., Slepak, T., Tachiiri, Y. and Simon, M. (1992) *Proc. Natl Acad. Sci. USA, 89*, 8794.
8. Coffey, A.J., Roberts, R.G., Green, E.D., Cole, C.G., Butler, R., Anand, R., Giannelli, F. and Bently, D.R. (1992) *Genomics, 12*, 474.

9. Green, E.D., Riethman, H.C., Dutchik, D.E. and Olson, M.V. (1991) *Genomics,* **11,** 658.
10. Abbott, C., West, L., Povey, S., Jeremiah, S., Murad, Z., Discipio, R. and Fey, G. (1989) *Genomics,* **4,** 606.
11. Saxon, P.J., Srivatsan, E.S., Leipzig, G.V., Sameshima, J.G. and Stanbridge, E.J. (1985) *Mol. Cell Biol.,* **5,** 140.
12. Schmid, C.W. and Jelinek, W.R. (1982) *Science,* **216,** 1065.
13. Nelson, D.L., Ledbetter, S.A., Corbo, L., Victoria, M.F., Ramírez-Solis, R., Webster, T.D., Ledbetter, D.H. and Caskey, C.T. (1989) *Proc. Natl Acad. Sci. USA,* **86,** 6686.
14. Irwin, M.H., Moffatt, R.J. and Pinkert, C.A. (1996) *Nature Biotech.,* **14,** 1146.
15. Ludecke, H.-J., Senger, G., Claussen, U. and Horsthemke, B. (1989) *Nature,* **338,** 348.
16. Johnson, D.H. (1990) *Genomics,* **6,** 243.
17. Han, J., Lu, C.-M., Brown, G.B. and Rado, T.A. (1991) *Proc. Natl Acad. Sci. USA,* **88,** 335.

9 Fingerprinting

9.1 Random amplified polymorphic DNA (RAPD)

RAPD, also known as arbitrarily primed PCR, allows the detection of polymorphisms without prior knowledge of nucleotide sequence. The polymorphisms may be used as genetic markers and may also be used for the construction of genetic maps. The method utilizes short (c. 10 nucleotide) primers of arbitrary nucleotide sequence that are annealed in the first few cycles of PCR at low stringency. The low stringency of the early cycles ensures the generation of products by allowing priming with mismatches between primers and template. The subsequent PCR cycles are performed at a higher stringency after the generation of some initial products that now have ends complementary to the primers (see *Figure 9.1*). Alternatively, an intermediate stringency primer annealing step may be used throughout the PCR to achieve the same outcome. RAPD is also a technique ideally suited to fingerprinting applications because it is fast, requires little material and is technically easy. It should be noted, however, that some workers have experienced the production of nonparental PCR bands in the offspring of known pedigrees. For this reason RAPD would not be the method of choice for applications such as human paternity testing and pedigree analysis, where absolutely unequivocal results are essential. Where results are not required to be as stringent, RAPD certainly has an important role. In this context, RAPD has been used to fingerprint strains of serovars of *Bacillus thuringiensis*, the most commonly used biological insecticide. Similarly, the technique has been used to examine clinically important strains of other bacterial species (see Section 12.3). Plants are particularly suited to RAPD analysis. Thus the technique has (a)　　5'$ACTGTGTCAATC$3'　　RAPD Primer

(b)　　　5'$ACTGTGTCAATC$3'
　　　3'TCTGGTACAATTAGGACAGTCATGATCGT→5'

　　　　　5'$ACTGTGTCAATC$3'
　　　3'GGCGACATGGTTAGAAGCACCGTAGTCGA→5'

　　　　　5'$ACTGTGTCAATC$3'
　　　3'AGAACCACAGTTAGCGCGTACGTAAAGCT→5'

　　　　　5'$ACTGTGTCAATC$3'
　　　3'CATGACGTCATTACCACAGCCCCGTTAAT→5'

(c)　　　5'$ACTGTGTCAATC$CTGTCAGTACTAGCA→3'
　　　3'TGACACAGTTAGGACAGTCATGATCGT→5'

　　　　　5'$ACTGTGTCAATC$TTCGTGGCATCAGCT→3'
　　　3'TGACACAGTTAGAAGCACCGTAGTCGA→5'

　　　　　5'$ACTGTGTCAATC$GCGCATGCATTTCGA→3'
　　　3'TGACACAGTTAGCGCGTACGTAAAGCT→5'

　　　　　5'$ACTGTGTCAATC$GTGTCGGGGCAATTA→3'
　　　3'TGACACAGTTAGCACAGCCCCGTTAAT→5'

FIGURE 9.1: (a) *An arbitrary sequence RAPD primer.* **(b)** *Mismatched annealing of primer to partially complementary sequences within total DNA sample under low-stringency conditions.* **(c)** *RAPD primer incorporated into PCR products – bottom strands are now fully complementary to the primer – PCR continues under higher stringency annealing conditions.*

been used extensively in plant breeding studies, and applications to strawberries, wheat, oat, barley, soya bean, tomato, potato and corn have all been reported. A small leaf disk taken using a standard paper punch from a seedling's cotyledon can be extracted to provide sufficient material for 20 RAPD analyses [1]. A typical RAPD result examining maize inbred lines is shown in *Figure 9.2*. For additional RAPD applications see Sections 12.1 and 12.3.

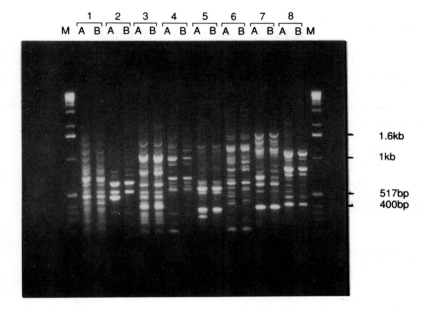

FIGURE 9.2: *The use of RAPD to characterize near-isogenic maize inbred lines. 10 ng of genomic DNA from maize near-isogenic lines A and B was subjected to PCR, using eight randomly chosen primers and the following cycling conditions: 25 sec 95°C, 35 sec 45°C, 120 sec 73°C for 40 cycles. Samples were analyzed on a 1.5% agarose gel; M = size markers (1 kb ladder; Gibco-BRL). Note that, as expected, the majority of primers produce an identical pattern with both the A and B lines; however, primers 2 and 5 produce an extra band in the inbred line A.*

9.2 Amplified fragment length polymorphisms (AFLPs)

The AFLP technique [2] is based on the selective PCR amplification of restriction fragments from a total restriction enzyme digest of genomic DNA. After restriction enzyme digestion of the DNA, oligonucleotide adapters are ligated to the restriction fragments. The selective amplification is achieved using amplimers designed from both the restriction enzyme recognition sequence and the adapter sequence, extending a short way (about three bases) into the restriction fragments themselves. Only those primers that match the restriction fragment at their 3′ end are amplified if a 3′-exonuclease deficient thermostable DNA polymerase is used; analogous to an ARMS reaction (see Section 11.1). Many restriction fragments are amplified simultaneously, thus sequencing gels are employed to ensure adequate resolution of the products. A typical AFLP analysis is shown in *Figure 9.3*. This analysis is of particular interest because it also demonstrates the technique of pooling phenotypically similar samples (bulk segregant analysis) [3], to identify markers that are linked to specific traits, in the case shown in *Figure 9.3*, white or red beet. Here, presumably the bands shown by arrows, represent markers that are tightly linked to the gene responsible for the red coloration.

9.3 Microsatellites

Microsatellites, otherwise known as TG repeats or CA repeats, are stretches of approximately 10–60 repeats of the dinucleotide. The repeating TG units on one DNA strand will be complementary to repeating CA units on the other strand. It is the variable number of these repeat units that confers polymorphism to microsatellites. There are up to 10^5 TG microsatellites interspersed through the human genome and, unlike the related minisatellite repeats or variable number tandem repeats (VNTRs), they do not tend to cluster towards the telomeres of the chromosomes. Microsatellites are not confined to the human or primate genomes. There are at least twice as many microsatellites in the mouse genome compared to the human, and there is a considerable allelic difference between inbred strains. For this reason, dinucleotide repeats should allow the construction of a high-resolution map of the mouse genome using recombinant inbred strains. Genetic relationships in birds have also been studied using microsatellites. In addition to TG repeats, TC repeats have also been identified in barn swallows and the pied flycatcher. Furthermore, these microsatellites can be amplified in other bird species [4].

TG microsatellites may be found by hybridizing an oligonucleotide (oligo AC) to a Southern blot of restriction-enzyme digested DNAs from genomic clones, isolating hybridizing fragments and subcloning these. The sequenced subclones then provide genomic DNA sequence flanking the microsatellite, from which PCR primers may be designed. Primers are designed according to the principles described in Section 2.4.1; in addition, any complementarity to Alu repeat

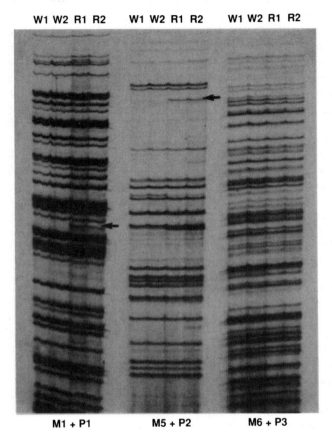

W1 W2 R1 R2   W1 W2 R1 R2   W1 W2 R1 R2

M1 + P1   M5 + P2   M6 + P3

FIGURE 9.3: *Autoradiograph of an AFLP analysis of sugar-beet DNA. Sugar-beet plants were separated into two pools of two according to the phenotypic red or white coloring within the root tissue. Standard AFLP amplifications were carried out on 1 µg of DNA from each of the four pooled samples using the primer combinations M1 + P1, M5 + P2 and M6 + P3. Primers M1, M5 and M6 were 5'-end labeled using T⁴ polynucleotide kinase and [γ-³³P]ATP (see Section 2.4.3) and the AFLP products were electrophoresed on a 6% polyacrylamide sequencing gel which was dried and autoradiographed. The arrows indicate bands present in the red pools (R1 and R2) and absent in the white pools (W1 and W2).*

sequences (see Section 8.5) should be avoided. Given the nature of the allelic variation of TG microsatellites, the primers should also be situated as close to the microsatellite as possible. This improves subsequent electrophoretic resolution of the PCR products and simplifies the analysis of the results.

Fingerprinting genomic DNAs using microsatellites is usually carried out by incorporating a radiolabel into the PCR product. This may be achieved either by incorporating the radiolabel at the 5′ end of one of the primers (see Section 2.4.3) or, alternatively, by incorporating a radiolabeled dNTP into the PCR product (see Section 2.3.1). If the latter option is adopted, the choice is usually to incorporate [α-^{32}P]dCTP in order to label the AC strand to high specific activity. The alternative is not to radiolabel the PCR product but to silver stain the gel after denaturing polyacrylamide gel electrophoresis. If the PCR products are radiolabeled, they are electrophoresed similarly and the gel subsequently autoradiographed. A phenomenon of the PCR of dinucleotide repeat microsatellites is the production of 'shadow' bands, hence only one strand is radiolabeled if end-labeling is chosen. The 'shadow' bands arise from slippage of the polymerase on the template strand during PCR due to the repetitive nature of the template sequence. Some workers have advocated reducing the number of PCR cycles and the primer concentration (to about 0.2 µM) to overcome this potential problem. A typical result showing microsatellite fingerprinting and the segregation of alleles is shown in *Figure 9.4*.

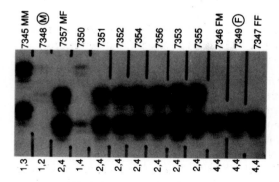

K1345 Family

FIGURE 9.4: *Microsatellite typing of a three-generation family (K 1345) at the D21S167 locus. The figure shows an autoradiograph of an acrylamide gel used to separate the fragments produced by PCR using primers that flank a CA repeat at this locus. M, mother; F, father; MM, maternal grandmother; MF, maternal grandfather; FM, paternal grandmother; FF, paternal grandfather. The alleles present in each family member are shown below the lanes.*

9.4 Variable number tandem repeats (VNTRs)

Variable number tandem repeat (VNTR) markers are similar to the microsatellite markers discussed in Section 9.3 except that the repeating units are larger. VNTRs also tend to cluster towards the telomeres of the chromosomes rather than being interspersed throughout the genome. However, these markers are particularly useful in determining identity and establishing the relatedness between individuals. An individual's genotype is obtained by determining the number of repeat units for each allele at each VNTR locus examined. One VNTR locus, D1S80, has a marker repeat of 16 bp with allele sizes from 350 to 1000 bp (14 to >50 repeat units). The D1S80 marker alone provides a discrimination power of 95–98%. The D1S80 marker is detected using the AmpliFLP™ D1S80 PCR amplification kit (P.E. Applied Biosystems). An associated product is the AmpliFLP™ D1S80 allelic ladder, which contains all alleles between 16 and 41 repeats, serving as a reference for calibration.

9.5 Human leukocyte antigen (HLA) class II typing

Human leukocyte antigen molecules are a diverse group of cell-surface glycoproteins involved in the regulation of immune responses. All class II cell-surface antigens comprise an α- and a β-chain. The genes encoding these chains are found on chromosome 6 and they are organized into three regions, DR, DQ and DP. Each region encodes one α-chain, while DQ and DP also encode one β-chain, and DR also encodes several β-chains (see *Figure 9.5*). There is extensive genetic polymorphism at the HLA class II loci, for instance the DRB1 locus has over 40 alleles. Furthermore, these are mainly localized to the second exon of the gene. Therefore, most class II sequence variation can be analyzed with a pair of PCR primers to conserved regions flanking this exon.

Characterization of the allelic variation of the HLA loci is important in the following areas:

(i) matching donors and recipients for bone and other tissue transplants to avoid rejection after transplantations;
(ii) as a marker system for paternity testing;

HLA Class II Genes

FIGURE 9.5: *Map of the HLA class II region. The α-chain loci are shown as open boxes, the β-chain loci as filled boxes. Unexpressed loci are shown as stippled boxes. The locus nomenclature is according to reference [5].*

(iii) individual identification or elimination in forensics;
(iv) susceptibility to a range of autoimmune diseases, where characterization of specific alleles or haplotypes can provide an early opportunity for therapy;
(v) examining evolutionary relationships.

The actual typing of HLA alleles can be performed using the techniques described in Chapter 11. The choice of technique depends on the application of the typing; however, the nature of the HLA polymorphisms described above lends itself to analysis by dot blots or reverse dot blots (see Sections 11.2 and 11.3) and kits are commercially available for this. Alternatively, RFLP analysis (see Section 11.4) can be performed after PCR with several primer pairs [6]. ARMS (see Section 11.1) is also an ideal method for analyzing HLA polymorphism [7,8], as is direct sequencing (see Chapter 7) [9]; however, sequencing cannot determine the phase of two polymorphisms, which is important in determining specific alleles; but this can be achieved using double ARMS [10]. Because of the preferential amplification of specific alleles, ARMS can be particularly useful for HLA analysis in forensic applications when the biological evidence material may comprise a mixture of genotypes. This type of sample is common in the analysis of vaginal or anal swabs from rape cases. Here, most of the genetic material is derived from the victim's epithelial cells, with only trace amounts of DNA originating from the sperm cells of the rapist. Another application, the tissue-typing for donor–recipient matching for transplants, often has to be performed quickly; the PCR-based typing techniques discussed here have facilitated this and greatly improved success and survival rates after transplant surgery.

References

1. Deragon, J.-M. and Landry, B.S. (1992) *PCR Methods Appl.,* **1,** 175.
2. Vos, P., Hogers, R., Bleeker, M., Reijans, M., van de Lee, T., Hornes, M., Frijters, A., Pot, J., Peleman, J., Kuiper, M. and Zabeau, M. (1995) *Nucleic Acids Res.,* **23,** 4407.
3. Michelmore, R.W., Paran, I. and Kesseli, R.V. (1991) *Proc. Natl Acad. Sci. USA,* **88,** 9828.
4. Ellegren, H. (1992) *Auk,* **109,** 886.
5. Nomenclature for Factors of the HLA System (1988) *Immunogenetics,* **28,** 391.
6. Westman, P., Kuismin, T., Partanen, J. and Koskimies, S. (1993) *Eur. J. Immunogenet.,* **20,** 103.
7. Fernandezvina, M., Moreas, M.E. and Stastny, P. (1991) *Hum. Immunol.,* **30,** 60.
8. Browning, M.J., Krausa, P., Rowan, A., Bicknell, D.C., Bodmer, J.G. and Bodmer, W.F. (1993) *Proc. Natl Acad. Sci. USA,* **90,** 2842.
9. Gyllensten, U. and Erlich, H.A. (1988) *Proc. Natl Acad. Sci. USA,* **85,** 7652.
10. Lo, Y.-M.D., Patel, P., Newton, C.R., Markham, A.F., Fleming, K.A. and Wainscoat, J.S. (1991) *Nucleic Acids Res.,* **19,** 3561.

10 Characterizing Unknown Mutations

In this chapter we describe techniques that analyze for single base differences between otherwise identical PCR products. These techniques are susceptible to the replication errors that can occur in PCR; it is therefore important to observe the guidelines described in Chapter 3 to avoid contamination and maximize specificity. It is particularly important that any erroneous sequence should never become a significant proportion of the PCR product. This situation should not arise if an adequate concentration of genomic DNA is used in the PCR ($\geq 10^4$ molecules or 0.1 ng of human genomic DNA).

10.1 Denaturing gradient gel electrophoresis (DGGE)

DGGE, followed by ethidium bromide or silver staining, is a technique that is capable of detecting a large proportion of polymorphisms in genomic DNA that has been amplified by PCR. The technique is based on the fact that DNA heteroduplexes differing by a single base pair have slightly different melting characteristics. The transition from double- to single-stranded DNA can be monitored by polyacrylamide gel electrophoresis by virtue of the reduced electrophoretic mobility of single-stranded DNA. Therefore when two similar heteroduplexes are electrophoresed through a denaturing gradient (which may be a chemical denaturant, such as formamide or urea, or a thermal gradient) DNA melting will occur at slightly different positions within the gradient, at which point the migration rate through the gel is slowed down, allowing resolution of the original similar heteroduplexes. Clearly, although this technique allows the determination of differences between two samples, it does not characterize the differences. Characterization of polymorphisms and mutations has to be performed using other methods, as described in Chapter 7.

DNA molecules melt in discrete segments that are dependent on DNA sequence and base composition. These segments, known as melting domains, have their own melting temperatures and they melt sequentially on passage through the denaturing gradient. DGGE can detect single base differences in all but the final domain. For this reason an artificial final domain is usually added during PCR. This is achieved by adding around 40 G and C residues to one of the PCR primers. This GC-rich region, which is added to one end of the PCR product, is known as a GC-clamp. The effect on the melting characteristics of a DNA fragment after adding a GC-clamp is shown in *Figure 10.1*.

The practical aspects of gradient gel preparation, equipment and electrophoresis procedures are described fully by Myers *et al.* [1]

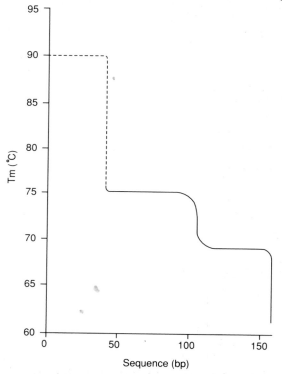

FIGURE 10.1: *The melting characteristics of a hypothetical DNA fragment representing the melting map predicted from DNA sequence. The x-axis represents the nucleotide sequence of the fragment, including a GC-clamp (broken line). Along the y-axis is the melting temperature (T_m) calculated at each base-pair position. The DNA sequence will be denatured at temperatures above the line and remain as duplex DNA at temperatures below the line. The fragment melts in three domains: 90°C for the GC-clamp, and 75°C and 70°C at about 100 and 150 base pairs, respectively.*

10.1.1 Parallel DGGE

This aspect of DGGE comprises the PCR amplification of a large genomic fragment (a GC-clamp is not necessary here) which is subsequently digested with one or more frequent-cutting restriction enzymes. Several digested PCR products from different individuals are then electrophoresed in adjacent lanes on separate denaturing gradients (e.g. 10–50% and 40–80% denaturant) as shown in *Figure 10.2*. Polymorphisms are then identified by bands present in some gel lanes that are not present in others, or by a mobility shift in some lanes with respect to the others. A typical experimental result using parallel DGGE is shown in *Figure 10.3*.

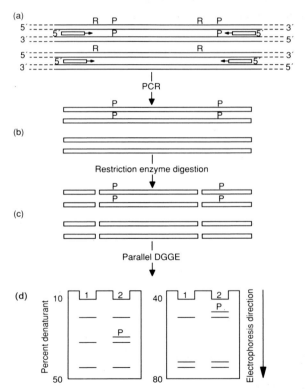

FIGURE 10.2: *Parallel DGGE and identifying polymorphisms.* **(a)** *Two genomic DNA duplexes with polymorphisms (P) and restriction enzyme recognition sites (R).* **(b)** *The PCR products.* **(c)** *The DNA fragments after restriction enzyme digestion.* **(d)** *The restriction enzyme digests are electrophoresed on two denaturing gradient gels with different denaturant concentration gradients. The extra band in lane 2 of each gel indicates the presence of a polymorphism. Polymorphisms associated with smaller restriction fragments are seen on the left-hand gel, those associated with larger fragments appear on the right-hand gel.*

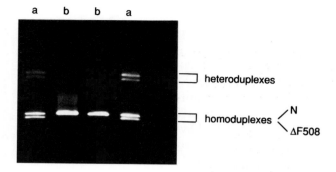

FIGURE 10.3: *Ethidium bromide stained parallel DGGE gel (10–60% denaturant) of PCR products from an N/ΔF508 heterozygote (a) and a normal exon 10 sample (b) amplified with CFTR gene exon 10 primers. The gel was run at 160 V for 5 h (N denotes a normal allele; ΔF508 is the most common cystic fibrosis mutation in Caucasian populations).*

10.1.2 Perpendicular DGGE

Perpendicular DGGE comprises co-electrophoresing two samples through a denaturing gradient gel perpendicular to the gradient. The combined PCR products are loaded across the entire width of the gel and, after electrophoresis, a melting curve for the DNA samples is generated. Where the denaturant concentration is low, the DNA migrates according to its molecular weight. Where the denaturant concentration is high, the fragments remain near the top of the gel due to the formation of a partially single-stranded molecule, as shown in *Figure 10.4*. Perpendicular DGGE is particularly useful for determining the melting domain structure of specific DNA fragments for the optimization of gradients for parallel DGGE. As described for parallel DGGE, the PCR-amplified DNA can be cut using frequent-cutting restriction enzymes; if the original PCR DNA substrate was from a single individual then the melting curves from any DNA fragment that split are allelic. Subsequent sequence analysis then allows characterization of the polymorphism. A result of perpendicular DGGE is shown in *Figure 10.5*.

10.2 Single-strand conformation polymorphism (SSCP)

SSCP [2] is another gel electrophoresis method that allows the detection (but not characterization) of mutations and polymorphisms.

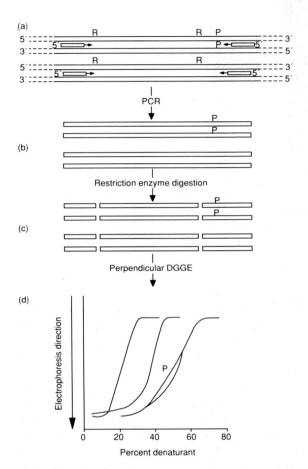

FIGURE 10.4: Perpendicular DGGE and identifying polymorphisms.
(a) Two genomic DNA duplexes with a polymorphism (P) in a
heterozygous individual and restriction enzyme recognition sites (R).
(b) The PCR products of each allele. **(c)** The DNA fragments after
restriction enzyme digestion. **(d)** The restriction enzyme digest is
electrophoresed on a denaturing gradient gel perpendicular to the
gradient. Three melting curves are seen, one for each restriction fragment.
The third melting curve exhibits two lines of transition, indicating the
presence of a polymorphism associated with that restriction fragment. To
characterize the polymorphism the DNA can be isolated from the gel
where the lines of transition have separated. This DNA can be PCR
amplified again and directly sequenced to compare the sequence of one
allele to the other.

(a)

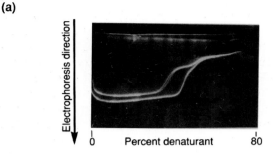

(b)

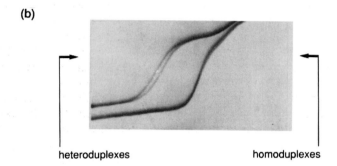

heteroduplexes   homoduplexes

FIGURE 10.5: **(a)** *Ethidium bromide-stained perpendicular DGGE gel (0–80% denaturant) of a DNA sample from an N/ΔF508 heterozygote amplified with CFTR gene exon 10 primers. 400 µl of PCR product was loaded on the gel.* **(b)** *An enlarged negative image of* **(a)**.

SSCP analysis is based on the principle that the electrophoretic mobility of a molecule within a gel matrix is sensitive to the size, charge and shape of the molecule. Under nondenaturing, or native conditions, single-stranded DNA has a folded structure that is imposed by intramolecular interactions dictated by its sequence. The technique differs from DGGE in that PCR products are denatured prior to electrophoresis in a nondenaturing gel as opposed to nondenatured PCR products being electrophoresed on a denaturing gel. In SSCP analysis a single nucleotide difference between two similar sequences is sufficient to alter the folded structure of one relative to the other. This conformational change is detected as a mobility difference upon gel electrophoresis. The sensitivity of the method allows detection of most single nucleotide differences in a fragment composed of several hundred nucleotides. Absolute characterization of mutations or polymorphisms can be performed by eluting appropriate bands from SSCP gels, reamplifying the eluted DNA and sequencing using one of the procedures described in Chapter 7.

SSCP analysis generally requires that the PCR products are labeled, and ideally of 400 base pairs or less. Labeling of the PCR products is not required under some circumstances, for example when mini gels (e.g. Phast gels, Pharmacia LKB) are used in conjunction with silver staining after electrophoresis. When labeling is required, this may be either via a radiolabel or via a fluorophore attached to the PCR primer(s) (see reference [3] and Section 2.4.3). Fluorescence-based SSCP is a recent innovation to the method and requires particularly sophisticated detection apparatus. One benefit of the fluorescence technology is the conversion of the technique from a qualitative to a quantitative assay [3]. Radiolabeling can be performed using $[\gamma\text{-}^{32}P]$ATP and polynucleotide kinase, as discussed in Section 2.4.3, or by the direct incorporation of α-radiolabeled dNTPs during the PCR. If large PCR products are analyzed, the labeling method of choice is the direct incorporation of radiolabeled dNTPs during the PCR. This then allows the PCR product to be cleaved with restriction enzymes to reduce individual fragments to ≤ 400 bp, each of which is radiolabeled. If polymorphisms are present in the initial PCR product, they will be more easily detected in the appropriate fragments generated by the restriction enzyme digest. The general scheme for SSCP is shown in *Figure 10.6*.

A critical aspect of SSCP analysis is uniformity throughout electrophoresis. This is because the conformation that a given single-stranded DNA molecule will adopt is dependent on temperature, and ion and solvent concentrations. In particular, temperature rises should be avoided during the electrophoretic run. To achieve this, thin gels that carry less current, water jacketed electrophoresis tanks and metal plates, which act as heat sinks, attached to one side of the gel have all been used.

The addition of glycerol to 5–10% in SSCP gels has been shown to improve the sensitivity of the method. Furthermore, commonly used acrylamide concentrations are in the region of 5% with 1–2% cross linking. With these conditions, approximately 90% of polymorphisms will be detected in PCR products of up to 450 bp, and approximately 99% of polymorphisms will be detected in PCR products of up to 300 base pairs. An SSCP result is shown in *Figure 10.7*.

10.3 Chemical cleavage of mismatches

The chemical cleavage of mismatches (CCM) method [4] is based on the fact that specific reagents can react with mispaired bases of DNA.

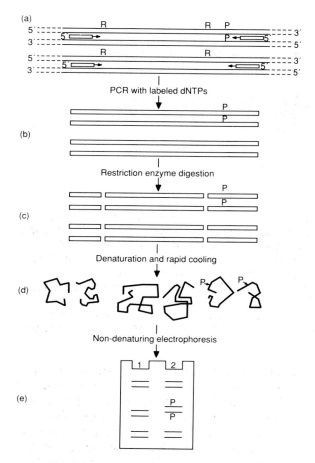

FIGURE 10.6: *SSCP and identifying polymorphisms. (a) Two genomic DNA duplexes with a polymorphism (P) and restriction enzyme recognition sites (R). (b) The PCR products. (c) The DNA fragments after restriction enzyme digestion. (d) After heat denaturation and rapid cooling each DNA single strand takes on a conformation unique to its nucleotide sequence. (e) The DNA single strands are electrophoresed on a nondenaturing gel. The mobility within the gel is dependent on the conformation adopted by each single strand. The mobility shift of the two bands in lane 2 of the schematic gel are indicative of polymorphisms.*

The subsequently modified bases make the DNA cleavable with piperidine. Allelic differences in DNA can therefore be studied using this method. Thus, screening for unknown mutations can be carried out using CCM. PCR products from 'normal' (probe) and 'mutant' (test) DNAs are radiolabeled as for SSCP (see Section 10.2). The two PCR products are combined, denatured and reannealed. This process

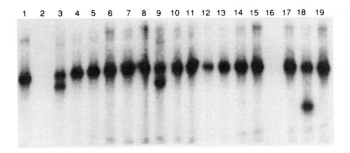

FIGURE 10.7: *SSCP analysis of exon 11 of the CFTR gene. Radiolabel was incorporated into PCR products by the addition of [α-³²P]dCTP into the PCR. Unlabeled dCTP was reduced to 125 µM. After denaturation of the PCR products in 95% formamide, 10 mM NaOH, 0.05% xylene cyanol and 0.05% bromophenol blue, the samples were electrophoresed on a 5% acrylamide 5% glycerol gel. After electrophoresis the gel was fixed, dried and autoradiographed. Lane 1, normal control; lane 2, space; lane 3, G551D mutation; lane 9, 1717-1 G→A mutation; lane 16, space; lane 18, R560T mutation; lanes 4, 5, 6, 7, 8, 10, 11, 12, 13, 14, 15, 17 and 19, normal exon 11.*

reproduces the two original homoduplexes and creates two new heteroduplexes, each containing one strand from each of the original PCR products. Sequence differences between the original DNAs will therefore exist as mismatches in the new heteroduplexes. Osmium tetroxide reacts with mismatched T residues and hydroxylamine reacts with similarly mismatched C residues. Treatment of the mismatched heteroduplexes with these chemicals followed by piperidine treatment cleaves the appropriate DNA strand at the mismatched position. Given that C and T residues are reactive with these chemicals, all mutations are detectable since mispaired A and G will be converted to T and C mismatches in the opposite sense strand.

When the chemically reacted heteroduplexes are electrophoresed on denaturing polyacrylamide gels and autoradiographed, bands appear from the radiolabeled fragments after chemical cleavage (*Figure 10.8*). Therefore, if a sizing sample, such as a DNA sequencing reaction, is electrophoresed on the same gel, it becomes possible to determine the approximate region of the PCR product that exhibited the original mutation. The particular chemical used in the base modification reaction will also help pinpoint the type of mutation. Nevertheless, unequivocal characterization of the mutation must be carried out by sequencing the PCR product(s) (see Chapter 7).

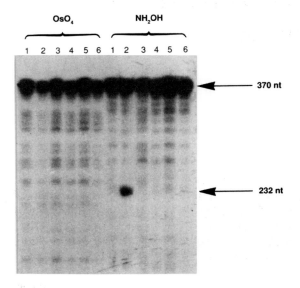

FIGURE 10.8: Chemical mismatch analysis of a 370 bp section of the dystrophin gene transcript (spanning exons 38–40) amplified from lymphocyte RNA by RT PCR (see Section 5.3). The samples are from five Duchenne muscular dystrophy patients (tracks 1–5) and a control plasmid (track 6). Patient 2 yields a cleavage fragment of approximately 230 nt with hydroxylamine. On sequencing the product it was found that the patient had a C to T transition at nucleotide 5759, resulting in the nonsense mutation Gln_{1851} to Stop [5]. The mismatch band arises because the C residue 232 nt from the 5′ end of the probe (normal) DNA is mispaired against an A residue in the heteroduplex formed with the antisense strand of the target (mutant) DNA.

References

1. Myers, R.M., Sheffield, V.C. and Cox, D.R. (1988) in *Genome Analysis: A Practical Approach* (K. Davies, ed.). Oxford University Press, Oxford, p. 95.
2. Orita, M., Iwahana, H., Kanazawa, H., Hayashi, K. and Sekiya, T. (1989) *Proc. Natl Acad. Sci. USA,* **86,** 2766.
3. Makino, R., Yazyu, H., Kishimoto, Y., Sekiya, T. and Hayashi, K. (1992) *PCR Methods Appl.,* **2,** 10.
4. Cotton, R.G.H., Rodrigues, N.R. and Campbell, R.D. (1988) *Proc. Natl Acad. Sci. USA,* **85,** 4397.
5. Roberts, R.G., Bobrow, M. and Bentley, D.R. (1992) *Proc. Natl Acad. Sci. USA,* **89,** 2331.

11 Analysis of Known Mutations

There are several PCR-based approaches to the analysis of known mutations. Only those routinely used are described in detail and Section 11.6 summarizes infrequently used methods of mutational analysis. Their relative advantages are compared and contrasted. For further details of these methods the reader is encouraged to refer to the respective seminal references cited.

In the analysis of mutations for medical diagnostic purposes, unequivocal results are essential. It is therefore recommended that all precautions to avoid PCR contamination are taken, as detailed in Chapter 3. In the event of equivocal results it is important to carry out a secondary analysis; ideally direct sequencing of amplified DNA (see Section 7.1) should be performed. This is because the result may be due to the presence of a rare or uncharacterized mutation, polymorphism or even deletions or rearrangements in regions primed either by amplification primers or hybridization oligonucleotides for the PCR product.

11.1 The amplification refractory mutation system (ARMS)

11.1.1 Basic principles

A typical ARMS assay [1] comprises two PCRs, each conducted using the same substrate DNA. ARMS relies on one primer of a pair of PCR primers being specific for one allele. In a second PCR reaction a primer specific for the other allelic variant is used. The allele specificity of these primers is conferred by the 3′ nucleotide of the primer which complements one allele but not the other. This specificity is maintained by the absence of a 3′ to 5′ proofreading activity in *Taq* DNA polymerase. Some other thermostable DNA

(a)

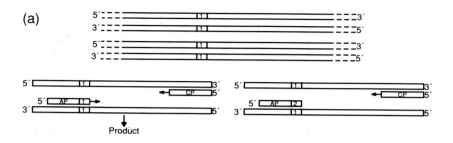

(b)

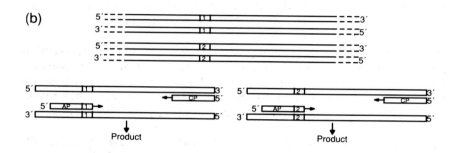

(c)

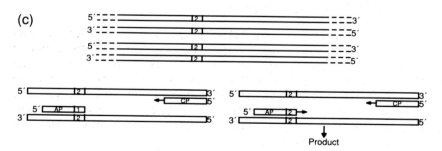

FIGURE 11.1: *The ARMS assay:* **(a)** *normal homozygote DNA (allele 1);* **(b)** *heterozygote DNA (alleles 1 and 2); and* **(c)** *affected homozygote DNA (allele 2). In* **(a)** *only the normal allele is present therefore only the primer specific to allele 1 is incorporated into the PCR product with this substrate DNA. In* **(b)** *both alleles are present, therefore in their respective PCR reactions, both ARMS primers give rise to PCR product. In* **(c)**, *with only allele 2 present, only the allele 2 primer has a complementary substrate and is extended and generates amplified product. AP = ARMS primer; CP = common primer.*

polymerases, such as Vent™ and *Pfu* DNA polymerases, which do possess a 3′ to 5′ proofreading activity (see Section 2.2), cannot therefore be used for this application. ARMS also requires that the enzyme's ability to initiate primed synthesis from a mismatch is severely impaired such that amplification is essentially nonexistent. The scheme of the ARMS assay is shown in *Figure 11.1*.

It is good practice to co-amplify a different region of the genome with a pair of internal control primers to ensure that the PCR has been efficient in each reaction, thus helping to avoid false negative results. Inspection of the PCR products by agarose gel electrophoresis and ethidium bromide staining is the only subsequent detection required to determine which allele(s) is present in the DNA sample under investigation. Therefore additional enzymatic manipulation steps, hybridizations or the use of radioisotopes are not needed.

The ARMS reaction has been applied successfully to the analysis of an increasingly large number of polymorphisms, germ-line mutations and somatic mutations. ARMS has been applied to carrier detection and prenatal diagnosis of inherited disease and in the detection of residual disease during and after cancer therapy (see Section 11.7).

ARMS has also been described in the literature as allele-specific PCR (ASP), PCR amplification of specific alleles (PASA) and allele-specific amplification (ASA).

11.1.2 ARMS primer design

When designing primers for an ARMS assay, the genomic sequence as opposed to the cDNA sequence must be considered. If the genomic DNA sequence is not known but the intron/exon boundaries within cDNA are characterized, it will usually be possible to position the common primer and ARMS primers so as to generate a suitable product for agarose gel electrophoresis. If this is not the case, then some intron sequence will be required. This can be obtained using the methods described in Chapter 8.

Although an ARMS test may be designed using primers in the region of 20 nucleotides in length, it is useful to increase the length of the primers to about 30 nucleotide residues. When these longer primers are used the possibility of establishing a generic set of reaction and thermal cycling conditions increases. Destabilizing the allele-specific primers by introducing deliberate mismatches close to the 3′-terminal nucleotide often improves specificity, but this may reduce yield. When introducing additional mismatches, one must consider the position within the primer, the GC content of the five or six nucleotides preceding the 3′ nucleotide and the discriminatory 3′ nucleotide that is dictated by the difference between the alleles and the type of mismatch. The nearer to the 3′ terminus of the primer that a destabilizing mismatch is incorporated, the greater is the effect on destabilization. A qualitative ranking of the destabilizing effect of additional mismatches is CC > CT > GG = AA = AC > GT. The effect of additional mismatches on the ARMS assay has to be determined empirically; however, it should be noted that the choice of control amplification fragment occasionally affects the specificity of the ARMS primers.

If ARMS is used to detect a frame-shift mutation, no further destabilization is required. This is because the deletion or insertion can provide additional mismatching between one primer and the nonrespective allele. An example of good and poor primer design for the Tay–Sachs exon 11 insertion mutation is shown in *Figure 11.2*.

The design of the common primer is less demanding than for the allele-specific primers. In general, the common primer should be selected for a region that has approximately 50% GC content, should not share 3′ complementarity with either allele-specific primer or

(a)
```
Normal allele        5'- ACCTGAACCGTATATCCTATGGCCCTGACTGGAAGGATTTCTA- 3'
Normal ARMS primer       GGCATATAGGATACCGGGACTGACCTTCCT- 5'
Mutant ARMS primer       tatAgATAGGATACCGGGACTGACCTTCCT- 5'
```

(b)
```
Exon 11 insertion allele  5'- ACCTGAACCGTATATCTATCCTATGGCCCTGACTGGAAGGATTTCTA- 3'
Normal ARMS primer            ggcAtATAGGATACCGGGACTGACCTTCCT- 5'
Mutant ARMS primer            TATAGATAGGATACCGGGACTGACCTTCCT- 5'
```

(c)
```
Normal allele        5'- ACCTGAACCGTATATCCTATGGCCCTGACTGGAAGGATTTCTA- 3'
Normal ARMS primer            TATAGGATACCGGGACTGACCTTCCTAAAG- 5'
Mutant ARMS primer           gATAGGATACCGGGACTGACCTTCCTAAAG- 5'
```

(d)
```
Exon 11 insertion allele  5'- ACCTGAACCGTATATCTATCCTATGGCCCTGACTGGAAGGATTTCTA- 3'
Normal ARMS primer            tATAGGATACCGGGACTGACCTTCCTAAAG- 5'
Mutant ARMS primer            GATAGGATACCGGGACTGACCTTCCTAAAG- 5'
```

FIGURE 11.2: *Primer design. (a) and (b) The normal ß-hexosaminidase A exon 11 allele and the mutant allele with the four base insert underlined, shown with well-designed primers that maximize mispairing between ARMS primers with their nonrespective allele. (c) and (d) The same alleles shown with poorly designed primers that have only one mismatched base when nonrespective alleles are primed. Mismatched bases are shown in lower case and bold.*

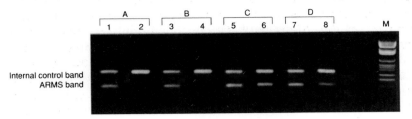

FIGURE 11.3: ARMS analysis of the normal ß-hexosaminidase A exon 11 allele and the mutant allele with the four-base insert (see Figure 11.2). The primers used were the 'well-designed' primers. Odd-numbered lanes contain products of the common and normal primers; even-numbered lanes contain products from the common and mutant primers. Subject A is a normal control, patients B and C are Tay–Sachs carriers (patient B is normal at this allele and therefore carries a different mutation). Patient D is affected by Tay–Sachs disease (patient D has one normal exon 11 allele and therefore both chromosomes have different mutations).

internal control primers, should provide an appropriate-sized fragment for gel electrophoresis and should not have repeated or unusual sequences. A typical ARMS assay result is shown in *Figure 11.3.*

11.2 Allele-specific oligonucleotide (ASO) hybridization to immobilized PCR products – dot blots

This technique involves the immobilization and denaturation of PCR products to a membrane, such as nitrocellulose or nylon, followed by hybridization with ASOs [2], the 'dot blot'. Using carefully derived conditions, an oligonucleotide that is fully complementary to one allele, an 'ASO', will hybridize to that allele only. The other allelic variant will, similarly, only hybridize with its specific oligonucleotide probe. The detection of hybridization of the ASO is via a signal-generating label linked to the ASO, such as a radiolabel, fluorophore or biotin. These labels are detected, respectively, by autoradiography, fluorescence after the appropriate wavelength UV irradiation and conjugation to an avidin– or streptavidin–enzyme conjugate that will generate a colorimetric or chemiluminescent signal. To avoid false negative results with this method it is essential to confirm that the PCR reaction was efficient by the examination of an aliquot of the reaction mixture by gel electrophoresis.

The choice of either the sense or antisense strand as target for the ASOs may affect their specificity; if the mutation or polymorphism confers a G–T mismatch between one allele and the ASO for the alternative allele, then it may not be possible to discriminate between the alleles. In these circumstances the alternative strand should be the target for the ASOs, where a C–A mismatch would be made.

An ASO dot-blot test is usually restricted to the analysis of a single pair of alleles by virtue of the signal generation system chosen. Furthermore, it is impractical to establish a generic protocol for ASO hybridization since the hybridization conditions are always dependent on the melting temperature (T_m) of the ASO and its complementary sequence, which will vary from one mutation or polymorphism to another. For these reasons the reverse dot blot (see Section 11.3) was developed.

11.3 Reverse dot blots

Unlike the dot blot, the reverse dot blot allows the simultaneous analysis of several alleles in a single test. The method is similar to the dot blot except that the format is inverted. ASOs that are tailed with poly(dT) in an enzyme reaction using terminal transferase and dTTP are immobilized on to a membrane by UV irradiation [3]. Hybridization is then carried out using a biotinylated or radiolabeled PCR product as the probe. The main advantage with respect to the dot blot is that several probes can be immobilized to detect specific alleles from a single PCR in which several primer pairs were combined. The principal limitations of the reverse dot blot are as for the dot blot probed with ASOs. Specifically, because the T_m for each immobilized ASO will be allele-dependent (for a constant-length ASO) each filter can only have bound ASOs with a similar T_m. Optimal ASOs usually comprise approximately 20 nucleotides if they are around 50% GC in content. Therefore, if altering the T_m of an ASO is attempted by adjusting its length, the probe specificity will also be affected.

As with ASO-probed dot blots, hybridization analysis must be performed post-PCR. Detection of hybridization is also required; the signal may, similarly, be a radiolabel if a ^{32}P-labeled deoxynucleoside triphosphate is incorporated into the PCR dNTP mix, fluorimetric if fluorophore-tagged PCR primers are used, or colorimetric or chemiluminescent if biotinylated PCR primers are used.

11.4 RFLP analysis

Traditionally, restriction fragment length polymorphism (RFLP) analysis was carried out using Southern blotting (Section 3.1) [4], but the technique may take up to a week to perform. Once an RFLP has been identified, it can now be detected by PCR rather than by Southern blotting, as long as some of the surrounding DNA sequence is known. If this is not the case, the surrounding sequence may be determined by cloning the restriction fragments detected by the probe and sequencing the cloned inserts. Alternatively, the probe sequence might be used in a vectorette PCR (see Section 8.1). The surrounding sequence, once determined, can then serve to design PCR primers. After PCR, the product is digested with the appropriate restriction enzyme and electrophoretic separation demonstrates the presence or absence of the restriction site. With this type of analysis it is useful to have a control for the restriction digestion in order to avoid a misdiagnosis. The simplest way to provide the control is to position the PCR primers such that the amplified target contains another nonpolymorphic copy of the restriction site, if one exists within range of the PCR. Alternatively, a copy of the restriction site may be incorporated into one of the PCR primers (see Section 4.1.1), or another target containing the restriction site can be co-amplified in the PCR.

11.5 Gene diagnosis using multiplex methods

Since several targets can be co-amplified in one PCR, it is possible to combine or 'multiplex' primer pairs in genetic diagnosis. Multiplex analyses can be exemplified in three main ways: the reverse dot blot (see Section 11.3); the mapping of gene deletions, for example in Duchenne muscular dystrophy (DMD); and the analysis of smaller deletions, frameshift and point mutations, where a variety of mutations give rise to a distinct clinical phenotype as in cystic fibrosis (CF).

11.5.1 Multiplex PCR analysis in Duchenne muscular dystrophy

DMD is an X-linked muscle-wasting disorder, and death from respiratory failure usually occurs by the age of 20. The defective gene is the dystrophin gene, which is exceptionally large, nearly 2400 kb, comprising about 70 exons. The majority of dystrophin gene mutations are deletions and most of these are found in two 'hot-spot' regions. In multiplex analysis, PCR primers flanking deletion-prone exons of the dystrophin gene are pooled for simultaneous amplification. Any of the regions in patient DNA that fail to amplify are identified by gel electrophoresis. Inclusion of the appropriate controls (i.e. normal male DNA positive control and a no DNA negative control) helps to avoid misdiagnoses.

This method can rapidly detect in the region of 98% of DMD deletions [5, 6] and provides a rapid diagnosis compared to the previous method of Southern blotting genomic DNA samples digested by a variety of restriction enzymes and probing with a range of probes for each of the deletion-prone exons. An example of a multiplex PCR DMD diagnosis is shown in *Figure 11.4*.

11.5.2 Multiplex ARMS in cystic fibrosis

Several ARMS reactions can be combined in the same pair of reaction tubes even when closely situated variant loci are to be examined. The multiplexing of ARMS can even be performed where individual PCR products overlap, as demonstrated in the development of a multiplex

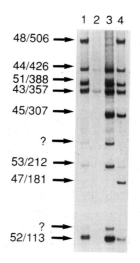

FIGURE 11.4: Multiplex PCR analysis of the Duchenne muscular dystrophy gene. PCRs are from DMD affected (samples 1, 2 and 3) and normal (sample 4) DNAs. Primers for exons 43, 44, 45, 47, 48, 51, 52 and 53 were multiplexed and PCR was carried out. The PCRs were electrophoresed on an acrylamide gel which was then silver stained. The exons and their amplicon size (bp) are indicated. ? = spurious PCR product. Sample 1, deleted exons 45 to 47; sample 2, deleted exons 45 to 53; sample 3, deleted exons 47 to 48; sample 4, normal.

ARMS test for the more common UK mutations of the CF transmembrane conductance regulator (CFTR) gene [7]. The multiplex reaction is essentially the same as individual ARMS reactions. Multiplex ARMS comprises two reactions as in conventional ARMS but each tube has a mix of primers for both normal and mutant alleles. Given that compound heterozygotes will have more normal alleles than mutant alleles per ARMS reaction tube when five or more loci are multiplexed, this distributes the number of expected amplification products between the two reactions and therefore obviates the requirement for an internal amplification control.

It should be noted that some ARMS primer sets demonstrate specificity when used in isolation; however, when multiplexed, their product yield may be very much reduced, particularly if deliberately introduced additional mismatches have been incorporated into the primers. In these situations it is necessary to revise any deliberate additional primer mismatches to compensate for this. Naturally, specificity of the revised primers will need to be confirmed. An example of a CF multiplex ARMS analysis is shown in *Figure 11.5*.

The reliability and accuracy of multiplex ARMS genetic testing has been demonstrated by the validation of the CF multiplex ARMS [7]. This validation study genotyped over 500 samples and the results were in complete agreement with the genotypes determined by other methods [7]. The ability to multiplex the ARMS analysis in one simple test has led to this becoming the method of choice in the analysis of known mutations in many genetic testing laboratories.

11.6 Other methods

11.6.1 Mutation detection by the introduction of restriction sites

This method of mutation or polymorphism detection [8] operates by incorporating a deliberate mismatch close to the 3′ terminus of a PCR primer, adjacent to a polymorphic nucleotide. After incorporation of the primer into a PCR product, the variant nucleotide of one allele matches the sequence of the primer to introduce a restriction enzyme recognition site. The corresponding nucleotide of the other allele does not complement the primer to form a restriction site. After restriction-enzyme digestion of the PCR product, cleavage will occur completely, partially or not at all, and this identifies which alleles were present in

(a)

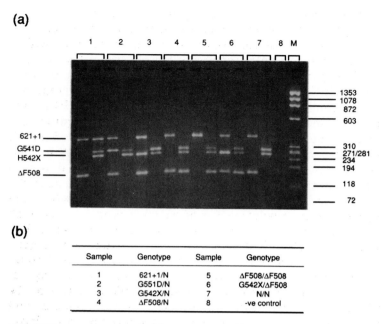

(b)

Sample	Genotype	Sample	Genotype
1	621+1/N	5	ΔF508/ΔF508
2	G551D/N	6	G542X/ΔF508
3	G542X/N	7	N/N
4	ΔF508/N	8	-ve control

FIGURE 11.5: Multiplex ARMS. **(a)** Gel analysis of the multiplex ARMS reactions for four of the mutations of the human CFTR gene from seven individuals (lane pairs 1–7). Lane 8 is a reagent blank (no DNA); M denotes size markers. The respective mutations detected in this assay are shown on the left of the panel. **(b)** The results of the multiplex ARMS assay in tabular form (N denotes a normal allele).

the template DNA. The limitation of the method involves the possibility of creating a restriction site within the primer in the confines of the genomic DNA sequence, since for some mutations there may not be a potential restriction site for incorporation. Also, the amplification of the target is likely to be compromised since the 3′ modification of the primer sequence is analogous to the property of ARMS primers, where 3′ mismatches are not extended. In addition, a failed restriction digest would imply homozygosity (only one allele present) for one allele, which may be incorrect. A partial restriction digest could also be misdiagnosed as heterozygosity (both alleles present). For these reasons it is desirable to incorporate a restriction site that occurs naturally within the PCR product and so would serve as a control for complete restriction-enzyme digestion.

11.6.2 Competitive oligonucleotide priming (COP)

Competitive oligonucleotide priming [9] employs allele-specific primers designed in the same manner as ASO probes. The allele specificity is derived from an appropriately variant nucleotide positioned roughly at the middle of each primer of a primer pair. The primer pairs are used simultaneously in a PCR in conjunction with a common primer, in a similar manner to an ARMS assay. However, in COP, the allele-specific primers must be distinguishable, for example one primer would be radiolabeled in one reaction; in a separate reaction the primer specific for the other allele would be radiolabeled. PCR is carried out and the products are analyzed by agarose gel electrophoresis. The expected result is an equivalent band in all samples analyzed. To discriminate between alleles, the gel is dried and autoradiographed. Where a band appears on the X-ray film, this corresponds to the incorporation of the radiolabeled primer into the PCR product. Thus, if the 'normal' primer were ^{32}P-labeled, the originating genomic DNA would either be from a homozygous 'normal' or heterozygous individual (assuming specificity of the primers). The second reaction would identify heterozygotes and homozygous affected individuals. In an improved format of COP [10] the individual primers of the primer pair may be conjugated to different signal-generation labels. For example, each allele-specific primer may be conjugated to a different fluorophore, such as fluorescein and rhodamine. After PCR, agarose gel electrophoresis and long-wavelength UV irradiation, the PCR product from heterozygous DNA will fluoresce yellow (rhodamine plus fluorescein incorporation). Homozygous DNA will fluoresce green (fluorescein-labeled primer incorporation) or red (rhodamine-labeled primer incorporation) for either allele, respectively. COP has rarely been the method of choice for known mutation analysis, given the difficulty of accurate heterozygote typing and the need for two detection systems (autoradiography in addition to gel electrophoresis) in the earlier format. The improved allelic discrimination contributed by positioning the allele-specific nucleotide at the 3′ terminus of primers as in ARMS allows ARMS to be performed as in the later COP format. Thus ARMS and COP combined have given rise to the fluorescent ARMS assay, which may be semi-automated and analyzed using an automated fluorescence-based DNA sequencer (P.E. Applied Biosystems ABI 373, ABI Prism™ 377 or Pharmacia ALF DNA-Sequencer™) (see Chapter 7).

11.6.3 Primer extension sequence test (PEST)

This method of mutation detection, like COP, relies on the extension of an allele-specific primer whose allele-specific nucleotide is within the primer. Thus the extension, or nonextension, of the primer relies on the hybridization of the primer to the template [11]. An ASO for the one allele is used in PCR after precision optimization, such that the ASO is extended only on the correctly matched template and generates a product with a downstream PCR primer. Another primer, upstream of the polymorphism/mutation to be detected and complementary to an invariant sequence, is also included. If the ASO does not hybridize, a PCR product is generated with the upstream primer. If the ASO does hybridize, it generates the PCR product itself. Heterozygous DNA templates are characterized by the formation of PCR products derived from both the upstream primer and the ASO. Since one primer is upstream of the other, products of different sizes are generated and may be resolved easily by electrophoresis. The drawbacks with this system are that only a single mutation can be analyzed per reaction and that each reaction must be optimized for the respective mutation, since the T_ms for different ASOs will vary. Specificity problems may also be encountered because of the noncompetitive annealing of the ASO to template DNA.

11.6.4 Mutation detection using *Taq* 5′ to 3′ exonuclease activity

Taq DNA polymerase does not possess a 3′ to 5′ exonuclease activity, but does exhibit a 5′ to 3′ exonuclease activity (see Section 2.2.1). These properties have been exploited to demonstrate the presence of a specific target DNA as the PCR proceeds [12]. This is achieved by including in the reaction another primer downstream of one of the conventional amplification primers. This additional primer is blocked for extension at the 3′ terminus and carries a label at the 5′ terminus. Extension from the primer upstream of the additional primer results in the detectable liberation of the 5′ label. The disadvantages of such a method for mutation detection are the inability to multiplex the reaction and the requirement for a separate reaction for each allele for each DNA sample. In addition, for this type of application the method would require the allele specificity to reside within the additional, 3′-blocked, 5′-labeled primer. This in turn would require this primer being designed in two forms, as for an ASO probe.

This technology (TaqMan™) has been adapted to an automated homogeneous assay [13, 14] which employs a fluorogenic probe. The probe has a reporter and a quencher dye attached. If there is extension from an allele-specific primer, the probe is cleaved by the 5′ to 3′ exonuclease activity of the polymerase, which releases the reporter dye and gives rise to increased fluorescence. The LS-50B 96-well microplate format luminescence spectrophotometer (P.E. Applied Biosystems) scans microplates in minutes. Alternatively, the P.E. Applied Biosystems PRISM™ 7700 Sequence Detection System can be used to provide a real-time data output as PCR progresses.

11.6.5 PCR amplification of multiple specific alleles (PAMSA)

This technique, like ARMS, utilizes the absence of a 3′ to 5′ proofreading activity in *Taq* DNA polymerase. The two techniques also share the basic allele discrimination conferred by the presence or absence of a matched 3′ end of an allele-specific primer. In PAMSA there is usually around a five nucleotide length difference between the primer designed for one allele and the primer specific for the other [15], but in the first report the lengths differed by up to 31 nucleotides [16]. Further improvements to the method, such as the introduction of additional mismatches near the 3′ end of primers for specificity improvement as with ARMS, have also been made [15]. This helps to maintain primer specificity at high template concentration. Intrinsic benefits are that one reaction is required and, in theory, specificity might be further enhanced since the allele-specific primers would be acting in competition with each other. Also in the PAMSA process there is an inherent internal control for the amplification reaction. In theory, PAMSA is amenable to multiplexing and could, with the appropriate choice of common primer for each allele-specific pair, be fractionated using sequencing gels and analyzed using automated DNA sequencers if fluorescent primers were employed.

11.7 Detecting residual disease after cancer therapy

The majority of cancers arise from a series of genetic alterations within a single cell. These mutations may be point mutations, gene deletions, gene rearrangements or gene amplifications. All of these alterations have been characterized using PCR. A particularly useful application of PCR is in the detection of chromosomal rearrangements associated with hematological malignancies. Some leukemias and lymphomas, such as chronic myelogenous leukemia (CML), are associated with specific chromosomal translocations. CML is associated with the aberrant 'Philadelphia chromosome' that occurs from a reciprocal translocation involving chromosomes 9 and 22 (t(9:22)). This translocation results in a transcriptionally active gene fusion that can be detected by RNA PCR (see Section 5.3). The mRNA from such gene fusions has been detected at a dilution of $1:10^5$ and also in remission samples from CML patients. The technique therefore allows the identification of CML patients in remission that are at high risk of relapse. Other translocations, such as those associated with many non-Hodgkin lymphomas, involve chromosomes

14 and 18 (t(14:18)), and the most common breakpoints occur in tightly clustered regions. In this situation it is possible to amplify across the breakpoint. Again, these breakpoints have been detected at a dilution of 1 : 10^5, thus PCR is capable of detecting the t(14:18) in minimal residual disease.

ARMS (see Section 11.1) is able to detect minimal residual disease because the discrimination of the technique allows the selective amplification of a mutant allele within a vast background of normal alleles. In some cancers, such as colorectal carcinomas [17] and pancreatic and hepatic neoplasms [18], there is a high frequency of mutations within specific codons of *ras* oncogenes. Studies are under way to examine whether mutations characterized in prediagnosed tumors can be detected in lymph nodes prior to metastasis, which may be useful in disease staging and as a prognostic indicator. ARMS has been applied in many such studies because of the ability to detect under-represented sequences. ARMS is also being evaluated in the detection of *ras* gene mutations in stool and colonic lavage samples, which could possibly lead to the earlier detection of colorectal carcinomas when there is a greater chance of successful treatment. The specificity and sensitivity of detection of one malignant cell in a background of 10^5 normal cells has been reported [19]. This selective amplification provides a simple downstream analysis, since after ARMS any background relative to amplified DNA becomes negligible. This can be compared with detecting a mutant allele in a background of normal alleles at the original ratio if conventional PCR followed by hybridization is employed. The use of PCR in cancer diagnostics is reviewed in reference [20].

11.8 Analysis of single cells for preimplantation diagnosis

From the discovery that the sensitivity of PCR allows amplification of DNA sequences from a single cell [21] came the concept that genetic disease diagnosis could be performed on human embryos produced by *in vitro* fertilization prior to implantation [22]. Early experiments showed that single blastomeres could be removed from human embryos at the six- or eight-cell stage and that highly repeated Y-chromosome sequences could then be detected by PCR (see Section 9.3). These experiments also showed that a large proportion of sampled embryos went on to develop normally to later stages *in vitro*

[23], and suggested that diagnosed embryos could be implanted in the mother's uterus and develop normally. These experiments have since been performed and normal births have resulted. Sexing of embryos in this way is particularly useful in the diagnosis of sex-linked inherited diseases.

Preimplantation diagnosis can even be made before fertilization. This requires analysis of DNA from the first polar body, which accompanies the oocyte. Such diagnoses are useful in inherited recessive diseases since a normal oocyte allele can be inferred if a mutant allele is found in the polar body. The oocyte could then be fertilized and implanted. However, the possibility of recombination should be considered if a disease-linked marker has been detected in the polar-body DNA, as there is then a finite risk of misdiagnosis if the mutation is not detected directly. For a review see reference [24].

References

1. Newton, C.R., Graham, A., Heptinstall, L.E., Powell, S.J., Summers, C., Kalsheker, N., Smith, J.C. and Markham, A.F. (1989) *Nucleic Acids Res.*, **17**, 2503.
2. Saiki, R.K., Bugawan, T.L., Horn, G.T., Mullis, K.B. and Erlich, H.A. (1986) *Nature,* **324**, 163.
3. Saiki, R.K., Walsh, P.S., Levenson, C.H. and Erlich, H.A. (1989) *Proc. Natl Acad. Sci. USA,* **86**, 6230.
4. Southern, E.M. (1975) *J. Mol. Biol.*, **98**, 503.
5. Chamberlain, J.S., Gibbs, R.A., Ranier, J.E., Nguyen, P.N. and Caskey, C.T. (1988) *Nucleic Acids Res.*, **16**, 11141.
6. Beggs, A.H., Koenig, M., Boyce, F.M. and Kunkel, L. (1990) *Human Genet.*, **86**, 45.
7. Ferrie, R.M., Schwarz, M.J., Robertson, N.H., Vaudin, S., Super, M., Malone, G. and Little, S. (1992) *Am. J. Human Genet.*, **51**, 251.
8. Haliassos, A., Chomel, J.C., Tesson, L., Baudis, M., Kruh, J., Kaplan, J.C. and Kitzis, A. (1989) *Nucleic Acids Res.*, **17**, 3606.
9. Gibbs, R.A., Nguyen, P.N. and Caskey, C.T. (1989) *Nucleic Acids Res.*, **17**, 2437.
10. Chehab, F.F. and Kan, Y.W. (1990) *Lancet*, **335**, 15.
11. Efremov, D.G., Dimovski, A.J., Janovic, L. and Efremov, G.D. (1991) *Acta Haematol.*, **85**, 66.
12. Holland, P.M., Abramson, R.D., Watson, R. and Gelfand, D.H. (1991) *Proc. Natl Acad. Sci. USA,* **88**, 7276.
13. Livak, K.J., Flood, S.J.A., Marmaro, J., Giusti, W. and Deetz, K. (1995) *PCR Methods Appl.*, **4**, 357.
14. Livak, K.J., Marmaro, J. and Todd, J.A. (1995) *Nature Genet.*, **9**, 341.
15. Okimoto, R. and Dodgson, J.B. (1996) *BioTechniques*, **21**, 20.
16. Dutton, C. and Sommer, S.S. (1991) *BioTechniques*, **11**, 700.
17. Span, M., Moerkerk, P.T.M., De Goeij, A.F.P.M. and Arends, J.W.A. (1996) *Int. J. Cancer (Pred. Oncol.),* **69**, 241.

18. Stork, P., Loda, M., Bosari, S., Wiley, B., Poppenhusen, K. and Wolfe, H. (1991) *Oncogene*, **6**, 857.
19. Billadeau, D., Blackstadt, M., Greipp, P., Kyle, R.A., Oken, M.M., Kay, N. and Van Ness, B. (1991) *Blood*, **78**, 3021.
20. Lyons, J. (1992) *Cancer*, **69**, 1527.
21. Li, H., Gyllensten, U., Cui, X., Saiki, R., Erlich, H. and Arnheim, N. (1988) *Nature*, **335**, 414.
22. Coutelle, C., Williams, C., Handyside, A., Hardy, K., Winston, R. and Williamson, R. (1989) *Br. Med. J.*, **229**, 22.
23. Handyside, A.H., Kontogianni, E.H., Hardy, K. and Winston, R.M.L. (1990) *Nature*, **344**, 768.
24. Verlinsky, Y., Handyside, A., Grifo, J., *et al.* (1994) *J. Assist. Reprod. Genet.*, **11**, 236.

12 Detecting Pathogens

PCR may be employed to detect a wide range of organisms, whether they are present in foodstuffs, the environment or biological or histological materials. The detection of pathogens is therefore an essential application of PCR with relevance to the food industry, environmental monitoring and medical, veterinary and botanical sciences.

The prime limitation to the use of PCR for such diagnostic applications is imposed by the amount of DNA sequence known for any given organism. Clearly, any DNA sequence upon which a diagnostic test is designed must be unique to the organism (or group of organisms). If the sequence is not unique, misdiagnosis is possible through false positive results. Furthermore, pathogen detection will often be performed against a significant background of host nucleic acid. Therefore, specificity for pathogen with respect to host is also important so as to reduce any background PCR products and maximize the sensitivity of the assay.

The main advantage of using PCR in the detection of pathogens is that a single cell or viral particle can be detected. Conventional pathogen characterization involving the culture of some organisms may take weeks, a PCR assay can be performed in less than an hour. Furthermore, some organisms cannot be grown *in vitro*, while others are difficult and slow to culture, therefore PCR has opened the way to detecting some organisms that were hitherto particularly difficult to detect. Furthermore, PCR has provided the means for easier sample collection for patients, such as small blood or urine samples or mouthwashes. These less invasive techniques, combined with the speed of PCR diagnosis, have greatly improved the service now offered to patients.

Amplicor™ kits (Roche Diagnostic Systems) have been developed to detect a variety of pathogens. These kits use the uracil *N*-glycosylase (Amperase™) protocol for avoiding carry-over contamination (see Section 3.2.1). The substitution of dUTP for dTTP in the PCR mix and pretreatment of all subsequent PCR mixes with uracil *N*-glycosylase prior to amplification specifically degrades dU-containing DNA while native DNA templates are unaffected. This is an effective measure to minimize the possibility of false positive diagnoses. Amplicor™ kits have also been adapted for automation using the Roche Cobas® Amplicor™ system. This automated device can be used for the detection of a variety of pathogens and further tests are under development. The instrument can also multiplex tests and perform them quantitatively. The clinical value of quantitative analyses is that it allows for the accurate monitoring of disease progression. Another recent development, from Johnson and Johnson Clinical Diagnostics, is the PCR pouch for the diagnosis of pathogens [1]. The system comprises a pouch formed from two sheets of heat-sealable plastic containing separate 'blister' compartments. Each 'blister' compartment contains different reagents. The test substrate is injected into a sample port connected to the first 'blister' containing the PCR reaction mix with biotinylated primers and complete with an anti-*Taq* DNA polymerase monoclonal antibody for automated hot start (see Section 3.5). The thermocycler for the pouch system has metal plates that sandwich the PCR 'blister' for temperature cycling at a rate of one cycle per minute. On completion of the thermal cycling the metal plates heat the reaction to denature amplicons, rollers then squeeze and burst the 'blister' forcing its contents along a tube. The rollers squeeze a second 'blister' containing streptavidin–horseradish peroxidase conjugate which binds biotinylated amplicons. The third 'blister' contains a wash solution, the fourth, a dye precursor that is converted by the horseradish peroxidase. The complete mixture is finally forced by the rollers to the detection area of the pouch with an array of oligonucleotide probes immobilized on to latex particles. Each spot of the array can have probes specific for different organisms, allowing the test to be multiplexed. The reagents are generic and generate a localized signal on any spot that successfully captured a biotinylated PCR product. The spots are automatically scanned to measure reflection density and provide the diagnosis. Because the pouch does not need to be opened after PCR for amplicon detection there is also no possibility of carry-over contamination to subsequent tests.

12.1 Fungi

There are limited literature examples of the use of PCR in characterizing human fungal pathogens. The reader should be aware, however, that this is an emerging application of PCR. Successes have been achieved using RAPD (see Section 9.1) in the characterization of the fungal pathogen *Aspergillus fumigatus*, the fungus responsible for invasive aspergillosis, an often fatal pneumonia in immuno-suppressed patients [2]. This fungus can also be detected using conventional PCR [3]. Progress has also been reported in the investigation of the epidemiology of *Candida albicans* and the diagnosis of systemic candidiasis [4]. Strain identification of the skin pathogen *Blastomyces dermatitidis* has also been achieved by RAPD analysis (see Section 9.1) [5].

PCR for fungal infection detection and/or strain identification is becoming increasingly important in the agricultural setting where it has been used to detect fungal pathogens specific to sugar-beet roots, potatoes, raspberry roots, strawberries and the common bean, amongst others. Soil can also be tested for the presence of specific fungi prior to planting. RAPD has been used in several instances for strain identification in association with distinct traits; for example, the relative aggressiveness of oil seed rape fungi has been studied and the degree of pathogenicity to various insect parasites and soil nematodes has been examined with the goal of developing strains of fungus for biological control. Similarly, rust-causing fungi are being evaluated for the biological control of weeds.

12.2 Viruses

12.2.1 Human retroviruses

Human retroviruses, for example human immunodeficiency virus types 1 and 2 (HIV-1 and -2) and human T-cell lymphotrophic virus types I and II (HTLV-I and -II) replicate through an RNA intermediate. The RNA intermediate can be detected after performing a reverse transcriptase step to generate a DNA template for PCR (see Section 5.3). However, during latency of infection, transcription is dormant. Nevertheless, latent infections by retroviruses can be detected by virtue of the retroviral replicative cycle (*Figure 12.1*). This cycle comprises conversion of the single-stranded RNA viral genome to a double-stranded circular proviral DNA molecule on entry into the host cell. The proviral DNA then integrates into the host cell DNA. Therefore latent infections can be distinguished from proliferating infections, allowing disease progression to be monitored. The distinction between latent and proliferating infection is made by the ability to detect viral RNA by RT PCR (see Section 5.3) in proliferating infection and viral DNA in both latent and proliferating infections. Quantitative PCR (see Section 13.1) also makes it possible to monitor the viral DNA load in infected individuals undergoing therapy [6].

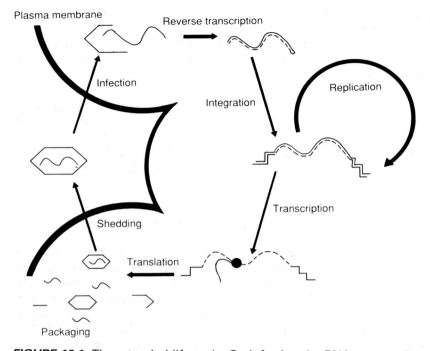

FIGURE 12.1: *The retroviral life cycle. On infection the RNA genome is released from the protein coat. Reverse transcription results in an RNA–DNA hybrid. The DNA then acts as template for the second DNA strand synthesis, which can then integrate into the genome of the host cell. Once integrated the virus is replicated along with the host cell. The infection may remain latent; however, the viral genome can also be transcribed, the RNA translated into protein which is then used to package viral RNA prior to the infectious viral particles being shed from the cell. Figure reproduced from Morgan and Darling (1993)* Animal Cell Culture, BIOS Scientific Publishers Ltd.

As a consequence of the replicative cycle and the reduced fidelity of the viral replicative polymerase and reverse transcriptase enzymes (compared to other polymerase enzymes), considerable heterogeneity between viral isolates is observed. For this reason it is necessary to target highly conserved regions of the viral genome for a PCR assay. These regions would comprise DNA sequences such as those coding for essential proteins such as the replicative enzymes, where conservation of sequence has been identified after DNA sequencing of a number of isolates. It may not be possible to define primers that target sequences which are completely nondivergent. If this is the case, sequences with minimal divergence should be chosen and the divergent bases substituted by inosine or the universal base (see Section 5.5) in the primers. Such base substitution would maintain equivalent priming efficiency, a particularly important consideration for quantitative PCR.

Amplicor™ HIV-1, HTLV-I and -II kits (Roche Diagnostic Systems) allow direct proviral detection. These tests are more accurate and sensitive than antibody tests or viral culture. Detection of these viruses for blood, blood products and organ/tissue transplantation is essential. PCR is also useful for the testing of neonates and in seroconversion.

12.2.2 Other clinically important viruses

Hepatitis B virus (HBV). HBV is a hepadna virus which also replicates through an RNA intermediate and can therefore be detected by amplification from RNA or DNA as the initial template nucleic acid. HBV infection is associated with 'serum hepatitis' and hepatocellular carcinomas, where HBV DNA has been found integrated into the cells' genomic DNA. PCR has improved the sensitivity of HBV detection over the existing serological methods, such that infections can now be confirmed in patients who are seronegative for all other HBV markers [7]. An Amplicor™ hepatitis C virus kit is available (Roche Diagnostic Systems) that can detect active viraemia in acute, chronic, resolved and relapsed stages. The kit uses reverse transcription of viral RNA to give a cDNA template. Both reverse transcription and PCR are performed using a single enzyme, *Tth* DNA polymerase.

Human papilloma virus (HPV). HPVs are small DNA viruses belonging to the papovavirus group; there are over 40 different types known. They are associated with a range of pathological manifestations, from warts and other benign tumors to some malignancies. In particular, distinct genital HPV types have been linked to cervical dysplasia and carcinoma (HPV types 16 and 18), while others (types 6 and 11) are associated with benign condylomas. Detection and typing of the various HPVs is essential in the understanding of their role in various diseases and cancer. PCR has been employed frequently in the analysis of HPV types in paraffin-embedded tissue (see Section 12.4.2) used in the elucidation of the association of specific types with the various disease manifestations (for example, see reference [8]). A recent PCR innovation, *in situ* PCR (see Section 12.5), has been applied to the detection of a single copy of HPV16 DNA in a human cervical carcinoma cell line [9].

12.3 Bacteria

Water contaminated by human fecal material is often the main means of transmission of diseases such as typhoid fever, dysentery and cholera. These diseases are caused by *Salmonella, Shigella* and *Vibrio cholerae,* respectively. Other waters in the environment may also serve as reservoirs for bacterial pathogens, such as the water in air-conditioning towers that may harbor *Legionella pneumophila*, estimated as being the causative agent in 85% of all legionella pneumonias (Legionnaire's disease). For these reasons environmental water must be monitored so that disinfection can be carried out before the possibility of a disease outbreak arises. PCR amplification of a target gene sequence followed by hybridization with specific gene probes provides both the sensitivity and specificity required for the monitoring procedure. The complete process of water sampling, recovery of bacteria from the sample, DNA extraction, PCR and hybridization has now been developed for a wide range of environmentally significant bacteria, with sensitivities of detection in the region of one cell per ml of water sample [10]. Similar environmental-monitoring PCR assays are being developed for soil pathogens such as *Bacillus anthracis*, the organism causing anthrax; here, the assay has another practical application in the inspection of slaughterhouses. Recently, tests developed using the random amplified polymorphic DNA (RAPD) technology (see Section 9.1) have been employed in the typing of different *Bacillus* species and serovars within these species [11]. RAPD is also being employed successfully in strain identification of the tuberculosis-causative agent, *Mycobacterium tuberculosis*, species and strain determination of *Staphylococcus* and strain identification of *Streptococcus pyogene*s.

The Enviroamp™ Legionella kit (P.E. Applied Biosystems) has been developed for the testing of environmental water samples. The kit detects the presence of *Legionella* bacteria and specifically identifies the presence of *L. pneumophila*. PCR is used to amplify the 5S ribosomal RNA gene from all members of the *Legionella* genus which is then detected by the use of *Legionella* 5S rRNA probes. *Legionella pneumophila* is identified by the amplification of the macrophage infectivity potentiator (*MIP*) gene followed by detection with a *MIP*-specific probe.

Mycobacterium tuberculosis can be detected using the Roche Diagnostic Systems Amplicor™ *Mycobacterium tuberculosis* kit. This is designed for the early and specific identification of the tuberculosis pathogen. The target sequence is a segment of the 16S ribosomal RNA gene common to all mycobacteria. Detection is by means of probes specific to *M. tuberculosis.*

The Amplicor™ *Chlamydia trachomatis* and *Neisseria gonorrhoeae* PCR detection kit is designed to detect both organisms if present in a sample prepared from urine or swabs. This is useful since there is an increasing incidence of co-infection with these organisms. The kit is available from Roche Diagnostic Systems.

Another adaptation has led to the ability to distinguish between a wide variety of bacterial genera and species. Here, a pair of primers for conserved regions of the adjacent 16S′ and 23S rRNA genes amplify spacer regions that show a high level of length and sequence polymorphism. One study showed that all 300 bacterial strains, belonging to eight genera and 28 species or serotypes, that were tested could be unambiguously typed by this method [12].

Yet another important application of PCR is in clinical microbiology, where the reaction is used in the detection of antibiotic resistance genes, such as methicillin resistance in *Staphylococcus aureus*, using primers specific for these genes. Not only does this procedure allow the appropriate treatment to be prescribed, but PCR can be employed subsequently to identify specific bacterial genotypes, which may be necessary in investigating the nature of antibiotic-resistant bacterial outbreaks in hospitals. This is possible by virtue of DNA repeats discovered in Gram-negative members of the family Enterobacteriaceae, for example. Primers specific for these variable repeat regions may then be used in applying the techniques described in Chapter 9.

12.4 Retrospective analyses using histological specimens

A vast archive of different tissue types and pathological preparations exists in hospitals and museums. This includes hematoxylin- and eosin-stained tissue sections, but the majority of specimens are formalin-fixed, paraffin-embedded tissues. DNA can be isolated from these samples and can also be amplified by PCR [13]. Recently, RNA PCRs have also been reported for formalin-fixed, paraffin-embedded archival material. It is therefore possible to study large numbers of patients and track infectious agents over 40 years old [8]. The fixative used and time of fixation affect the efficiency of subsequent PCR reactions [14], therefore if PCR is planned as part of a long-term study, the choice of fixative should be made taking this into consideration.

Although this section concentrates on the detection of pathogens, it should be noted that the analysis of histological material also allows the analysis of inherited and *de novo* mutations if genomic DNA is analyzed using techniques described in Chapter 11 or is amplified and sequenced directly using methods described in Chapter 7.

12.4.1 Hematoxylin- and eosin-stained sections

Hematoxylin- and eosin-stained sections are prepared from most blocks of tissue prepared for histological examination. Sections are cut from a paraffin-embedded block of tissue, de-waxed, stained and mounted on a microscope slide under a coverslip. The DNA isolation procedure consists of immersing the slide in xylene for 1 or 2 days to solubilize the resin holding the coverslip, scraping the section into a microfuge tube and incubating it with proteinase K for up to 5 days [15]. The sample is then deproteinized and ethanol precipitated prior to PCR analysis. The disadvantage of using hematoxylin- and eosin-stained sections is that the samples are destroyed, but these may be the only source of DNA if the original tissue block is not available.

12.4.2 Formalin-fixed, paraffin-embedded material

Formalin, a 4% aqueous solution of formaldehyde, is the most common histopathological fixative and, fortunately, it does not react with DNA unless the DNA is denatured. However, yields of DNA from formalin-fixed material depend on the time of fixation, due to leaching of the DNA during this process [16].

The DNA extraction procedure does not require de-waxing as for hematoxylin- and eosin-stained sections. Several sections are cut from the tissue block, these combined sections are then treated as described above for hematoxylin- and eosin-stained sections. DNA isolated in this way may not amplify efficiently on occasion, despite reasonable yields; if this is the case, dilution of the DNA usually overcomes this, suggesting that the problem is caused by a contaminant of the sample which inhibits *Taq* DNA polymerase.

12.4.3 Exfoliative cytology material

Exfoliative cytology material comprises cells collected by aspiration, washing or scraping, for example cells from the uterine cervix collected by cervical smear tests. If the number of cells is adequate, DNA may be extracted from these samples as described for hematoxylin- and eosin-stained sections. Where cell numbers are small, for instance in urine or pleural effusions, it may be necessary to remove the coverslip from the slide, paint a circle around the cell sample with nail varnish, to form a well, and then carry out the proteinase K digestion directly on the slide. After digestion, the sample can be pipetted into a microfuge tube and processed as before.

Archival cervical smear slides have been of great importance in the study of cervical neoplasia and epidemiological investigations, and have aided the association of particular HPV types with the respective pathological states (see Section 12.2.2) [7].

12.5 *In situ* PCR

This technique combines the extreme sensitivity of PCR with the cell-localizing capability of *in situ* hybridization. It is a technique that allows the detection of viral DNA, single-copy genes and gene rearrangements, or viral RNA and gene transcripts if incorporated with RT PCR (see Section 5.3). *In situ* PCR involves performing the reaction on fixed cells or tissue after semipermeabilizing the cell membranes to allow diffusion of the PCR reagents into the cells. Most workers advocate the 'hot-start' method (see Section 3.4) to enhance sensitivity and specificity; PCR is then initiated on a microscope slide beneath a sealed coverslip. The reaction is carried out with the slide placed on the heating block of the thermocycler and the detection of the PCR products is carried out either directly or indirectly. Direct detection of amplified sequences can be achieved if nonisotopic labeled nucleotides are incorporated into the amplified product, so allowing immunohistochemical detection. A less rapid, more sensitive but indirect detection method requires the additional step of *in situ* hybridization using amplicon-specific oligonucleotide probes. Both direct and indirect detection can also be employed; direct detection will show both specific and nonspecific amplified material, whereas indirect detection will show only specifically amplified material. This type of analysis might be important in establishing the optimum conditions for a routine test or screen which might ultimately be based only on direct detection.

The demanding nature of *in situ* PCR does bring associated complications and technical requirements for success. The major problems to consider are:

(i) The PCR process of thermal cycling should not destroy the cell morphology, so that it is still possible to associate specific cells or cellular components with the amplicons after they are detected.

(ii) The reagents for the reaction should be able to diffuse into the semipermeabilized cells.

(iii) The amplicons should not diffuse out of the semipermeabilized cells.

(iv) Tissue drying and loss of tissue adherence to the slide should not occur either during PCR or detection.

From the list above, it can be seen that the critical aspect of successful *in situ* PCR is the preparation of the cells or tissue and their fixation and permeabilization procedures. Long *et al.* [17] have carried out an extensive study and give exhaustive methods and comparisons between different sample types and between direct and indirect *in situ* PCR. RT *in situ* PCR is described in detail in reference [18].

The problems listed above have been overcome by the introduction of *in situ* PCR thermal cycling systems (e.g. Omnislide and SureSeal, Hybaid; *In-Situ* PCR System 1000, P.E. Applied Biosystems). Both of these examples incorporate a unique containment system for localizing the reaction components on to the tissue under investigation, together with an accurate and reliable instrument for thermally cycling the reaction *in situ* on the slides. Both instruments also provide accurate and reproducible thermal conditions for standard *in situ* hybridizations. The GeneAmp® *In-situ* PCR Core Kit (P.E. Applied Biosystems) has also led to the development of successful direct and indirect *in situ* PCR methods for the localization of viral and bacterial infections and the localization of specific gene sequences in formalin-fixed paraffin-embedded sections. With the direct incorporation of biotin into the PCR reaction or by hybridization after PCR using biotin-labeled probes followed by detection of products using biotin–streptavidin–alkaline phosphatase, the detection of, for example, human papilloma virus, cytomegalovirus, Epstein–Barr virus, *Helicobacter pylori* and Alu repeat sequences in formalin-fixed paraffin-embedded human tissue is possible. Detailed protocols may be found in reference [19].

References

1. Findlay, J.B., Atwood, S.M., Bergermeyer, L. *et al.* (1993) *Clin. Chem.*, **39**, 1927.
2. Loudon, K.W., Burnie, J.P., Coke, A.P. and Matthews, R.C. (1993) *J. Clin. Microbiol.*, **31**, 1117.
3. Spreadbury, C., Holden, D., Aufauvre-Brown, A., Bainbridge, B. and Cohen, J. (1993) *J. Clin. Microbiol.*, **31**, 615.

4. Hopfer, R.L., Walden, P., Setterquist, S. and Highsmith, W.E. (1993) *J. Med. Vet. Mycol.*, **31**, 65.
5. Yates-Siilata, K.E., Sander, D.M. and Keath, E. J. (1995) *J. Clin. Microbiol.*, **33**, 2171.
6. Piatak, M., Luk, K.-C., Williams, B. and Lifson, J.D. (1993) *BioTechniques*, **14**, 70.
7. Thiers, V., Nakahima, E., Kremsdorf, D., Mack, D., Schellekens, H., Driss, F., Goudeau, A., Wands, J., Sninsky, J. and Tiollais, P. (1988) *Lancet*, **2**, 1273.
8. Shibata, D.K., Arnheim, N. and Martin, W.J. (1988) *J. Exp. Med.*, **167**, 225.
9. Nuovo, G.J. (1992) *Amplifications*, no. 8, 1.
10. Atlas, R.M. and Bej, A.K. (1990) in *PCR Protocols: A Guide to Methods and Applications* (M.A. Innis, D.H. Gelfand, J.J. Sninsky and T.J. White, eds). Academic Press, San Diego, p. 399.
11. Brousseau, R., Saint-Onge, A., Préfontaine, G., Masson, L. and Cabana, J. (1993) *Appl. Environ. Microbiol.*, **59**, 114.
12. Jensen, M.A., Webster, J.A. and Straus, N. (1993) *Appl. Environ. Microbiol.*, **59**, 945.
13. Impraim, C.C., Saiki, R.K., Erlich, H.A. and Teplitz, R.L. (1987) *Biochem. Biophys. Res. Comm.*, **142**, 710.
14. Greer, C.E., Lund, J.K. and Manos, M.M. (1991) *PCR Methods Appl.*, **1**, 46.
15. Jackson, D.P., Hayden, J.D. and Quirke, P. (1991) in *PCR: A Practical Approach* (M.J. McPherson, P. Quirke and G.R. Taylor, eds). Oxford University Press, Oxford, p. 37.
16. Jackson, D.P., Lewis, F.A., Taylor, G.R., Boylston, A.W. and Quirke, P. (1990) *J. Clin. Pathol.*, **43**, 499.
17. Long, A.A., Komminoth, P., Lee, E. and Wolfe, H.J. (1993) *Histochemistry*, **99**, 151.
18. Nuovo, G.J., Gorgone, G.A., MacConnell, P., Margiotta, M. and Gorevic, P.D. (1992) *PCR Methods Appl.*, **2**, 117.
19. Nuovo, G.J. (1995) *PCR Methods Appl.*, **4**, S151.

13 Quantitative PCR

As discussed in the previous chapters, one of the main advantages of PCR is its high sensitivity, making it possible to detect rare DNA or RNA molecules. Furthermore, in the previous chapters, PCR has been examined in a qualitative mode, where the presence or absence of a product is indicative of the presence or absence of a target (assuming the appropriate controls have been included in the experiment). This chapter considers how PCR can be performed in a quantitative fashion for applications such as measuring gene dosage, gene expression and the extent of viral infections.

The exponential nature of PCR amplification lends it intrinsically to quantitative analysis. However, anything that may interfere with the exponential amplification will introduce errors. Therefore, for quantitative applications the PCR protocols should be tailored to minimize factors capable of affecting exponential amplification. Also, in this context, PCR is not wholly exponential since the reaction plateaus at around 10^8 copies of an amplicon. Therefore for quantitative work, PCRs should be maintained within about 20 cycles, during which the amplification is linear. With low numbers of starting target molecules, the amplification achieved after 20 cycles will require a sensitive detection system. However, with low numbers of starting molecules the linear range is extended over additional cycles. As a guide, one study has shown that the reaction is linear up to 30 cycles with 12–400 starting copies, up to 25 cycles with 200–3200 starting copies and up to 20 cycles with 3200–51 200 starting copies [1]. Ideally, the linear range of amplification for a given sample type should be defined as a prerequisite to quantitative PCR. This can be achieved using control samples or standards. Suitable standards should therefore be as close as possible to the composition of the target to be measured and ideally should use the same amplimers as the target DNA sequence. Standards should also be used at concentrations similar to those of the target, such that they have similar amplification kinetics and remain in the linear PCR range. For some target sequences, standards (also known as PCR mimics) are available commercially (see Appendix B). A schematic representation of the analysis of a competitive quantitative PCR is

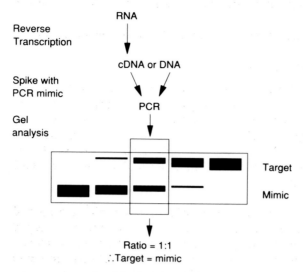

FIGURE 13.1: *Quantitative PCR using standards (or PCR mimics). PCRs carried out using DNA or cDNA as substrate can be spiked with varying amounts of mimic target which is amplified using the same PCR primers as the real target sequence. After gel electrophoresis the relative intensities of the target and mimic PCR products are compared. The concentration of the target is determined by extrapolation from the PCR mimic products.*

shown in *Figure 13.1.* Some elegant ways of overcoming the difficulties in achieving quantitative PCR have been devised and these are discussed below.

Detection and quantification of the amplified target sequences may be performed either indirectly or directly, but the trend is towards direct detection methods since a further intersample variable is eliminated. Radiolabeling of the amplicons, either via ^{32}P-end-labeled primers (see Section 2.4.3), or via the incorporation of an [α-^{32}P]dNTP into the PCR product (see Section 2.3.1) are arguably the most common methods of amplicon tagging. Of these alternatives, the use of ^{32}P-end-labeled primers is preferred since higher specific activities of product are achievable. This is because dNTP concentrations must be maintained to avoid interference with exponential amplification (i.e. the unlabeled dNTP content cannot be lowered to increase incorporation of the [α-^{32}P]dNTP).

The quantitation of the radiolabeled amplicons after gel electrophoresis can be accomplished in one of three ways:

(i) densitometric analysis of an autoradiograph of the electrophoretic gel;

(ii) scintillation counting of excised bands from the electrophoretic gel;

(iii) radioimaging of the electrophoretic gel.

The first option is the least favored. This is because the linear response range of X-ray film covers approximately three orders of magnitude only. The remaining two methods provide the dynamic range required for quantitation (*c.* five orders of magnitude). Radioimaging either with cross-wire ionization instruments or phosphorimaging devices is, however, simpler and more convenient than scintillation counting.

Much current research is aimed at examining nonisotopic tagging and detection. Some researchers are using chemiluminescent-tagged primers in conjunction with photoanalytical imaging systems. Alternatively, fluorophore tags may be used, with amplicons detected and quantified as described for real-time amplicon detection (see Section 3.1.2) and DNA sequencing (see Chapter 7).

With the TaqMan™ 5′ exonuclease detection system (see Section 11.6.4) together with the ABI PRISM™ 7700 Sequence Detection System (P.E. Applied Biosystems) thermal cycling and amplicon detection by using laser-induced fluorescence is performed concurrently. Thus, the increase in PCR products is measured as PCR proceeds. This instrument has a large linear dynamic range covering five orders of magnitude. Using the PRISM™ fluorescent labeling reagents several accumulating amplicons can be detected simultaneously. It is therefore possible to use one fluorophore with a mimic or reference target to give a standard curve and another with the amplicon under investigation. Because the detection is in real time it is also simple to ensure that measurements are made during the linear phase of amplification since readings are taken during each PCR cycle.

A comprehensive review of quantitative PCR may be found in reference [2].

13.1 DNA

For quantitative DNA PCR, the standards are initially amplified individually at a range of starting concentrations. From these data it is then possible to create a standard curve. This may be exemplified by considering the monitoring of the viral DNA load of HIV-infected patients receiving therapy (see Section 12.2.1). Suitable sources of HIV DNA for a DNA standard would be a cell line carrying a known amount of integrated viral genomes, ideally a single genome. Alternatively, a plasmid with an HIV DNA insert would be quite acceptable. For these types of DNA standards the initial quantitation of target DNA would be by cell counting and by OD_{260} readings, respectively. To simulate the real situation, the HIV DNA would ideally be diluted in peripheral blood mononuclear cells from an HIV-seronegative individual. In this example, overall quantitation would be improved by normalizing the HIV DNA load to the human genomic DNA load. The most commonly used standards to quantitate human DNA content have been the HLA genes (see Section 9.5) and the β-globin gene. This allows the co-amplification of a specific viral DNA sequence and a genomic DNA target. This improves the quantitative power of the assay, since any variation of the amplification efficiency would be observed in the yields of both amplicons, allowing normalization of the HIV amplicon to the genomic DNA amplicon.

13.2 RNA

Quantitation of RNA by PCR is a more challenging procedure than quantitation of DNA. This is because it is more difficult to provide a standard against which to measure the amplicon generated by RT PCR (see Section 5.3). Some studies have used internal or external DNA standards; however, the drawback with such standards is that it is not possible to assess the effects of intersample variation in the reverse transcriptase reaction. Therefore, only the DNA amplification reaction of RT PCR is controlled; while there is no control for the RT reaction. With RT reactions varying in efficiency from 5 to 90%, this is the biggest barrier to accuracy in quantitative RNA PCR. This problem has been addressed in some studies by the introduction of an external RNA standard. However, significant variation has been reported even between duplicate samples when using RNA standards. Another approach adopted by some groups has been to use an internal RNA standard. Here, calibration of the system is performed by co-amplifying an mRNA whose abundance does not vary, but this may not be feasible within the context of maintaining control and test targets at a similar concentration in the test sample. Another, widely used, standard for RNA quantitation is an internal synthetic standard that uses the same primer sequences as the target RNA but yields either a different-sized product or a similar-sized product that can be distinguished by restriction enzyme digestion. The disadvantage of this type of standard is that it cannot compensate for degradation of the target mRNA which would, to a large extent, be controlled if an internal mRNA standard were chosen. This is an important consideration given that RNA is more labile than DNA and that ribonucleases will be present in cell lysates.

References

1. Kellog, D.E., Sninsky, J.J. and Kwok, S. (1990) *Anal. Biochem.,* **189,** 202.
2. Ferre, F. (1992) *PCR Methods Appl.,* **2,** 1.

Appendix A

Glossary

Allele: one of the two or more different forms of a gene at a given locus.

Allele-specific oligonucleotide (ASO) probes: short, single-stranded oligonucleotide probes, differing in nucleotide sequence usually by only a single nucleotide, which recognize small differences in target DNA samples.

Amplicon: a PCR-amplified DNA fragment.

Amplification: increasing the number of copies of a specific DNA molecule in PCR.

Amplimer: an oligonucleotide that serves as a primer in PCR and defines the boundaries of an amplification product.

Anneal: the process of stringent hybridization of a single-stranded oligonucleotide to a single-stranded template nucleic acid to form a double-stranded, hydrogen bonded molecule. Annealing can take place between DNA and/or RNA.

Antisense: the antisense strand of protein-coding DNA is the strand complementary to the mRNA. *See also* Sense *and* mRNA.

Autoradiography: the process of exposing X-ray film to a radioactive source to generate an image that defines the position of the radioactive components relative to each other. The process generates an autoradiograph.

Base pair: a pair of complementary nucleotides which hydrogen bond the opposing strands of a DNA or RNA duplex or DNA/RNA heteroduplex.

Blunt end: an end of a duplex DNA fragment that has no overhang on one strand relative to the other (i.e. the strands are flush).

Bubble PCR: a synonym for chemical genetics. *See also* Chemical genetics.

cDNA: *see* Complementary DNA.

cDNA library: a collection of cDNA fragments, each cloned into a separate vector molecule.

Chemical genetics: a PCR technique that uses vectorette units to provide a primer annealing target on DNA of unknown sequence. *See also* Vectorette.

Chromosome: a structure of DNA and associated proteins in the nucleus of a eukaryotic cell or comprising the genome of a prokaryotic cell that contains the hereditary material of the cell in the form of a linear array of genes.

Cohesive end: the single-stranded end of a DNA fragment generated after digestion with a restriction enzyme that has a palindromic recognition site and cuts asymmetrically.

Complementarity: the specific binding of adenine to thymine (or uracil in RNA) and guanine to cytosine in opposite antiparallel strands of DNA or RNA.

Complementary DNA (cDNA): DNA synthesized by primer extension from mRNA by the use of a reverse transcriptase or certain DNA polymerases.

Concatemer: a larger DNA fragment formed by the ligation of multiple copies of a smaller DNA fragment usually generated by restriction enzyme digestion.

Contig: a stretch of DNA sequence built up from overlapping smaller segments of DNA sequence.

Cosmid: a genetically modified plasmid construct containing lambda bacteriophage DNA sequences that allow the insertion of very large pieces of DNA (up to 50 000 base pairs). Cosmids can be replicated in bacterial hosts.

Cross-hybridization: the binding of a DNA or oligonucleotide probe to a target other than its specific target, due to homology between the specific and nonspecific targets or lack of stringency in the hybridization conditions.

Denature: to cause the dissociation, chemically or thermally, of the complementary strands of a double-stranded nucleic acid; also known as melting.

Dideoxynucleotide: a deoxyribonucleotide analog that does not have a hydroxyl group at the 3′ position. Incorporation of a dideoxynucleotide into an enzymatically extending DNA strand blocks further extension of the strand, resulting in chain termination.

DNA: deoxyribonucleic acid, the single- or double-stranded helical molecule comprising the deoxyribonucleotides deoxyadenosine monophosphate, deoxyguanosine monophosphate (purine nucleotides), deoxycytidine monophosphate and deoxythymidine monophosphate (pyrimidine nucleotides).

Electrophoresis: a technique for separating molecules based on their differential mobility in an electric field. Agarose or polyacrylamide gels provide a simple medium for electrophoretic separation of nucleic acids.

Exon: one of two or more portions of a eukaryotic gene that comprises protein-coding sequence.

Exonuclease: an enzyme that degrades nucleic acids specifically from one end, either the 3′ end or the 5′ end.

Fidelity: the accuracy of a polymerase enzyme in synthesizing the complementary strand to the template nucleic acid.

Fluorescent: a fluorescent molecule is one that emits light at a specific range of wavelengths when hit by incident light of a shorter wavelength.

Fluorophore: a fluorescent molecule used to label oligonucleotides.

GC-clamp: a stretch of approximately 40 G and C residues added to the 5′ end of a PCR primer designed to raise the T_m of the end of the PCR product in which the primer, and therefore the clamp, is incorporated.

Gene: a sequence of nucleotides that codes for a protein product and which may be comprised of exons interspersed by introns.

Genome: the collection of hereditary components of an organism, contained in one set of the organism's chromosomes.

Genomic library: a collection of DNA fragments directly derived from an organism's genome that are each cloned into a separate vector molecule for subsequent replication.

Genotype: the collection of alleles that are present in an individual.

Haplotype: the genetic constitution of an individual with respect to one or more members of pairs of allelic genes.

Heterozygote: a cell or organism having two different alleles at a given locus on homologous chromosomes.

Homology: the similarity of nucleotide sequence between two distinct DNA molecules.

Homozygote: a cell or organism having the same allele at a given locus on homologous chromosomes.

Hybridization: the process of complementary base pairing between two single strands of nucleic acid.

Intron: a portion of a eukaryotic gene that does not code for protein although being transcribed into mRNA precursor molecules (i.e. occurs between two exons). *See also* Exon.

Kilobase: 1000 base pairs of DNA or 1000 bases of RNA.

Label: a molecule conjugated to, or radioisotope incorporated in, nucleic acid to permit detection of the nucleic acid or its complementary sequence after hybridization.

Library: a collection of different DNA fragments ligated into a vector and cloned into a bacterium or yeast, or just ligated to a vectorette. *See also* Vectorette.

Ligase: an enzyme that creates a phosphodiester bond between the 5′-PO_4 end of one polynucleotide and the 3′-OH end of another, to produce a single polynucleotide.

Ligate: to covalently join the ends of two or more DNA fragments.

Locus: the position on a chromosome where a gene or particular mutation is located.

Melt: *see* Denature.

Mimic: a PCR mimic is an amplicon of similar size and base composition to a PCR target amplicon, and is amplified using the same pair of PCR primers. It is added to a PCR at a known concentration at the start of the reaction and co-amplified with the target. The ratio of target to mimic PCR product is used to extrapolate the target to mimic ratio before the reaction.

mRNA, messenger RNA: an RNA molecule whose nucleotide sequence is translated into an amino acid sequence during protein synthesis. Mature eukaryotic mRNA molecules have a tract of enzymatically added adenosine residues (poly A) at their 3′ end; these are also known as poly(A)$^+$ mRNAs.

Mutation: a change in a gene that gives rise to a different protein product relative to that from the original species. Mutations are divided into point mutations, whose single base changes give rise to single amino acid changes in the protein, splicing mutations, which affect how the exons in mRNA are joined, and frameshift mutations caused by the insertion or deletion of base pairs in the gene which disturb the triplet coding register.

Northern blot: the RNA equivalent of a Southern blot. *See also* RNA *and* Southern blot.

Nucleoside: a nucleotide lacking a phosphate group. *See also* Nucleotide.

Nucleotide: the molecule comprising a purine (adenine or guanine) or pyrimidine (cytosine, thymine or uracil) base with a phosphate and a sugar, in turn comprising the fundamental monomeric structure of nucleic acids.

Oligonucleotide: a short stretch of single-stranded nucleic acid (usually synthetic DNA).

Palindromic: a DNA sequence on one DNA strand that is matched by the same complementary sequence on the other (e.g. 5′GAATTC3′ is complementary to 3′CTTAAG5′). *See also* Restriction enzyme.

PCR mimic: *see* Mimic.

Phosphoramidite: monomer units for the synthesis of oligonucleotides. Some specialty phosphoramidites allow the synthesis of modified or labeled primers.

Plasmid: an extrachromosomal, circular, self-replicating DNA molecule found in some bacteria. Most genetic manipulations are performed using plasmids.

Point mutation: *see* Mutation.

Poly(A)$^+$ RNA: *see* mRNA.

Polymerase: one of several enzymes that synthesize RNA and/or DNA from nucleoside or deoxynucleoside triphosphates, respectively, using a single-stranded nucleic acid as a template.

Polymerase chain reaction (PCR): an enzymatic DNA amplification method that comprises multiple rounds of primer extension by cycling template DNA and primers between temperatures that allow repeated DNA denaturation, primer annealing and primer extension.

Polymorphism: a variation in the sequence of DNA – not necessarily a mutation since it may occur within noncoding DNA or may be within a coding region but not affect the encoded protein due to the degeneracy of the genetic code. A polymorphism may also give rise to a variant protein not necessarily of clinical significance.

Primer: the oligonucleotide extended during primer extension. *See also* Amplimer *and* Primer extension.

Primer extension: the enzymatic extension of an oligonucleotide that has been annealed to a template, incorporating nucleotides complementary to the template. *See also* Template.

Probe: a fragment of DNA, RNA or an oligonucleotide that hybridizes to a complementary sequence of nucleotides in a single-stranded target nucleic acid.

Processivity: the number of nucleotides replicated by a DNA polymerase before the enzyme dissociates from the template.

Promoter: a sequence of DNA that RNA polymerase binds and uses to initiate transcription of the template into RNA.

Proofreading: the ability of a polymerase enzyme to excise incorrectly incorporated nucleotides during the synthesis of a strand complementary to the template molecule. Proofreading is accomplished by a 3′ to 5′ exonuclease activity associated with the polymerase. The proofreading ability contributes to the fidelity of a polymerase. *See also* Fidelity.

Proteinase K: an enzyme that degrades proteins and which is used in the isolation and purification of nucleic acids.

Recombinant: a DNA molecule generated by *in vitro* splicing of heterologous DNA molecules. This DNA is then reintroduced into a biological environment; a recombinant clone is a cell type transformed by a recombinant molecule.

Restriction enzyme: a bacterial enzyme that recognizes a short, specific DNA sequence and cleaves both DNA strands, usually within the recognition sequence. *See also* Palindromic.

Restriction fragment length polymorphism (RFLP): the variation in the fragment sizes when genomic DNAs from different individuals have been cleaved with the same restriction enzyme(s). The fragment size variation is the result of polymorphism between the DNA samples (*see also*

Polymorphism). An RFLP is 'informative' if a pattern of fragment sizes is associated with a particular mutation or disease. RFLPs can also be detected by PCR across the polymorphic site, followed by restriction enzyme digestion.

Reverse transcriptase: an enzyme capable of synthesizing single-stranded DNA from RNA in the 5′ to 3′ direction.

RNA: ribonucleic acid, the single-stranded molecule comprising the ribonucleotides adenosine monophosphate, guanosine monophosphate (purine nucleotides), cytidine monophosphate and uridine monophosphate (pyrimidine nucleotides). It is chemically closely related to DNA. RNA has a variety of functions; mRNA is the working copy of the genes and is translated into protein. Ribosomal and transfer RNA are components of the protein synthesis machinery. In addition, some RNA molecules have an enzymatic function.

Secondary structure: pertains to the structure formed by the folding of a DNA or RNA molecule.

Self-complementary: a DNA sequence of a single-stranded molecule that, on folding of the molecule, produces one or more regions of duplex DNA.

Sense: the sense strand of protein-coding DNA is the strand bearing the individual amino acid codons. *See also* Antisense.

Southern blot: DNA that has been electrophoretically separated and immobilized on to a solid support (nylon or nitrocellulose) in a denatured state for hybridization.

Sticky end: *see* Cohesive end.

Stringency: conditions that affect the specificity of hybridizations or annealing of two single-stranded nucleic acid molecules. Increasing the temperature and decreasing the ionic strength increase stringency and generally increase specificity. However, too stringent conditions will prevent even fully complementary sequences from annealing.

T_d: the dissociation temperature (°C). An approximation of the T_m (see T_m) for an oligonucleotide up to 25 nucleotides long. $T_d = 4 \times$ (number of G:C base pairs) $+ 2 \times$ (number of A:T base pairs).

Template: a single-stranded nucleic acid from which a complementary strand is synthesized by a polymerase.

T_m: the melting temperature (°C) at which the transition from double- to single-stranded DNA is 50% complete.

Transcript: an RNA molecule synthesized by an RNA polymerase. *See also* mRNA.

Transcription: the production of RNA molecules which are synthesized by an RNA polymerase.

Transformation: the introduction of a natural or cloned DNA molecule into an organism (usually a bacterium) so that the DNA molecule can be inherited by subsequent generations.

Translation: the production of a protein by conversion of the information contained in the mRNA molecule which determines the sequence of amino acids in the protein that is produced. *See also* mRNA.

Variable number of tandem repeats (VNTRs): copies or repeats of sequence motifs arranged in direct succession within a chromosome. The number of copies varies randomly at any locus from one unrelated individual to another.

Vector: a DNA molecule, usually of bacterial origin, such as a plasmid, phage or cosmid, that is used to clone genes or other DNA sequences of interest for introduction into bacterial or eukaryotic cells for propagation.

Vectorette: an oligonucleotide pair that forms a paired duplex only at the ends after the oligonucleotides are hybridized, therefore having a central mismatched region. One of the duplex portions is either blunt-ended or has a restriction enzyme cohesive end for ligation to restriction-enzyme-digested DNA. The vectorette provides the target for a PCR primer after it has been incorporated into a PCR product by primer extension from the other end of the PCR target. It is used in determining new DNA sequence adjacent to a region of known DNA sequence.

Wild-type: the normal allele of a gene.

Appendix B

Suppliers

Below is a list of the major suppliers of equipment, enzymes, reagents and consumables mentioned in the text.

The polymerase chain reaction is the subject of patent property assigned to Hoffmann-La Roche (see, for example, US-A-4176195 and US-A-4176202). When selecting materials and apparatus for performing the polymerase chain reaction we recommend that you refer to any specific instructions or caveats issued by the manufacturer.

Cloning vectors: Amersham International plc, Amersham North America, Boehringer Mannheim Biochemicals (BCL in the UK), CLONTECH Laboratories Inc. (UK distributors, Cambridge BioScience), Gibco-BRL Life Technologies, Invitrogen Corporation (UK distributor, R&D Systems Europe Ltd), Promega, Stratagene.

Consumables: Costar UK Ltd, Eppendorf, Gilson Inc., Midwest Scientific, P.E. Applied Biosystems, Stratagene, Techne.

Custom oligonucleotide synthesis: Appligene, Genosys Biotechnologies, CLONTECH Laboratories Inc. (UK distributor, Cambridge BioScience), Operon, P.E. Applied Biosystems, R&D Systems Europe Ltd.

DNA sequencing reagents and/or kits: Amersham, P.E. Applied Biosystems, Gibco-BRL Life Technologies, R&D Systems Europe Ltd.

Enzymes and biochemical reagents: Amersham International plc, Amersham North America, Boehringer Mannheim Biochemicals (BCL in the UK), Appligene, CLONTECH Laboratories Inc. (UK distributor, Cambridge BioScience), Gibco-BRL Life Technologies, New England Biolabs (UK distributor, CP Laboratories), 5 Prime → 3 Prime Inc. (UK distributor, CP Laboratories), Promega, Stratagene, United States Biochemical Corp. (UK distributor, Cambridge BioScience).

Gel electrophoresis equipment: Appligene, Gibco-BRL Life Technologies, Hoefer Pharmacia Biotech Inc., New England Biolabs (UK distributor, CP Laboratories), Stratagene.

Magnetic beads (streptavidin conjugate): Dynal Inc., Dynal (UK) Ltd.

mRNA isolation: CLONTECH Laboratories Inc. (UK distributors, Cambridge BioScience), Gibco-BRL Life Technologies, Invitrogen Corporation (UK distributor, R&D Systems Europe), Pharmacia (Biochrom) Ltd., Promega, Qiagen (UK distributor, Hybaid Ltd), Stratagene.

Nucleic acids and libraries: CLONTECH Laboratories Inc., Invitrogen Corporation (UK distributor, R&D Systems Europe), 5 Prime → 3 Prime Inc., Stratagene.

PCR mimics: CLONTECH Laboratories Inc. (UK distributor, Cambridge BioScience), Stratagene.

PCR product purification: P.E. Applied Biosystems, Promega, Qiagen (UK distributor, Hybaid Ltd), Stratagene.

Phosphoramidites and associated reagents: P.E. Applied Biosystems, Genosys Biotechnologies, Cruachem.

Thermal cyclers: Appligene, Biometra, Ericomp Ltd., Grant Instruments (Cambridge) Ltd., Hybaid, P. E. Applied Biosystems, 5 Prime → 3 Prime Inc. (UK distributor, CP Laboratories), M.J. Research Inc., Sanyo (UK distributor, Sanyo Gallenkamp plc), Stratagene, Techne.

Thermostable DNA polymerases: Amersham, Appligene, Boehringer Mannheim (Diagnostics and Biochemicals), CLONTECH, Flowgen Instruments Ltd, Gibco-BRL Life Technological, Hoefer Pharmacia Biotech Inc., Molecular Genetic Resources Inc., NBL Gene Sciences, New England Biolabs (UK distributor, CP Laboratories), P.E. Applied Biosystems, Pharmacia, Promega, Stratagene, Takara Shuzo Co. (UK distributor, Stratech Scientific Ltd; USA distributor, PanVera Corp.), United States Biochemical Corp. (UK distributor, Cambridge BioScience).

Ultraviolet products (transilluminators, safety equipment): Amersham, Anachem, Appligene, Fotodyne Inc. (UK distributor, TechGen International Ltd), Hoefer Pharmacia Biotech Inc.

Vectorette units for a variety of restriction enzyme cohesive or blunt ends: Genosys Biotechnologies.

Addresses

Amersham North America, 2636 South Clearbrook Drive, Arlington Heights, IL 60005, USA. Tel: 847 593 6300, 708 593 6300; fax: 874 437 1640, 708 593 8010.

Amersham International plc., Amersham Place, Little Chalfont, Bucks, HP7 9NA, UK. Tel: 0800 515 313, 01494 544 000; fax: 0800 616927, 01494 542 266.

Amicon Inc., 72 Cherry Hill Drive, Danvers, MA 01923, USA. Tel: 800 343 1397; fax: 508 777 6204.

Amicon, Upper Mill, Stonehouse, Gloucester, GL10 2BJ, UK. Tel: 01453 825 181; fax: 01453 826 686.

Anachem Ltd., Anachem House, 20 Charles Street, Luton, LU2 0EB, UK. Tel: 01582 456 666; fax: 01582 391 768.

Applied Biosystems (a division of Perkin Elmer), 850 Lincoln Center Drive, Foster City, CA 94404-1128, USA. Tel: 800 327 3002, 415 507 6667; fax: 415 638 998.

Applied Biosystems Ltd. (a division of Perkin-Elmer), Kelvin Close, Birchwood Science Park North, Warrington, Cheshire, WA3 7PB, UK. Tel: 01925 825 650; fax: 01925 828 196.

Appligene Inc., 1177-C Quarry Lane, Pleasanton, CA 94566, USA. Tel: 800 955 1274, 510 462 2232; fax: 510 462 6247.

Appligene, Pinetree Centre, Durham Road, Birtley, Chester-le-Street, Co. Durham, DH3 2TD, UK. Tel: 01914 920 022; fax: 01914 920 617.

Biometra Inc., 550 North Reo Street, Tampa, FL 33609-1013, USA. Tel: 813 287 5132; fax: 813 287 5163.

Biometra Ltd., Whatman House, St Leonard's Road, 20/20 Maidstone, Kent, ME16 0LS, UK. Tel: 01622 678 872; fax: 01622 752 774.

Bio-Rad Life Science Group, 2000 Alfred Nobel Drive, Hercules, CA 94547, USA. Tel: 800 4 BIORAD, 510 741 1000; fax: 800 879 222, 510 741 1060.

Bio-Rad Laboratories Ltd., Bio-Rad House, Maylands Avenue, Hemel Hempstead, Herts, HP2 7TD, UK. Tel: 0800 181 134, 01442 232 552; fax: 01442 259 118.

Bios Corporation, 291 Whitney Avenue, New Haven, CT, USA. Tel: 800 678 9487; fax: 203 562 9377. (UK distributor, Scotlab Ltd.)

Boehringer Mannheim, 9115 Hague Road, PO Box 50414, Indianapolis, IN 46250-0414, USA. Tel: 800 262 1640, 317 849 9350; fax: 800 428 2883, 317 576 2754.

Boehringer Mannheim (Diagnostics and Biochemicals) Ltd, Bell Lane, Lewes, East Sussex, BN7 1LG, UK. Tel: 0800 521 578, 01273 480 444; fax: 0800 181 087, 01273 480 266.

Cambridge BioScience, 25 Signet Court, Newmarket Road, Cambridge, CB5 8LA, UK. Tel: 01223 316 855; fax: 01223 60732.

CLONTECH Laboratories Inc., 1020 East Meadow Circle, Palo Alto, CA 94303, USA. Tel: 800 662 2566, 415 424 8222; fax: 800 424 1350, 415 424 1064. (UK distributor; Cambridge BioScience.)

Corning Costar UK Ltd, 10 The Valley Centre, Gordon Road, High Wycombe, Bucks, HP13 6EQ, UK. Tel: 01494 684 700; fax: 01494 464 891.

CP Laboratories, PO Box 22, Bishop's Stortford, Herts, CM23 3DX, UK. Tel: 01279 758 200; fax: 01279 755 785.

Cruachem Inc., 45150 Business Court, Ste. 550, Sterling, VA 22170, USA. Tel: 703 689 3390, 800 EASY-DNA; fax: 703 689 3392.

Cruachem Ltd, Todd Campus, West of Scotland Science Park, Acre Road, Glasgow G20 0UA, UK. Tel: 0141 945 0055; fax: 0141 946 6173.

Digene Corporation, 9000 Virginia Manor Road, Beltsville, MD 20705, USA. Tel: 800 344 3631, 301 470 6500; fax: 301 608 0696, 301 470 6798.

Dynal Inc., 5 Delaware Drive, Lake Success, NY 11042, USA. Tel: 800 638 9416; fax: 516 326 3298; email: techserv@dynalnsa.attmail.com.

Dynal (UK) Ltd, 10 Thursby Road, Croft Business Park, Bromborough, Wirral, Merseyside, L62 3PW, UK. Tel: 0151 346 1234; fax: 0151 346 1223.

Eppendorf–Netheler–Hinz GmbH, Biotech Products, Barkausenweg 1, Hamburg D-22339, Germany. Tel: 40 53801 0; fax: 40 53801 593. (UK distributor, Merck Ltd.).

Ericomp Inc., 6044 Cornerstone Court West, San Diego, CA 92121, USA. Tel: 619 457 1888; fax: 619 457 2937; email: sales@ericomp.com.

Flowgen Instruments Ltd, Lynn Lane, Shenstone, Nr Lichfield, Staffordshire, WS14 0EE, UK. Tel: 01543 483 054; fax: 01543 483 055.

FMC BioProducts, 191 Thomaston Street, Rockland, ME 04841, USA. Tel: 800 341 157, 207 594 3400; fax: 207 594 3491. (UK distributor, Flowgen Instruments Ltd.)

Genosys Biotechnologies, London Road, Pampisford, Cambs, CB2 4EF, UK. Tel: 01223 839 000; fax: 01223 839 200.

Gibco-BRL, *see* Life Technologies.

Gilson Inc., Box 27, 3000 West Beltline Highway, Middleton, WI 53562-0027, USA. Tel: 608 836 1551; fax: 608 831 4451. (UK distributor, Anachem Ltd.)

Glen Research, 44901 Falcon Place, Sterling, VA 20166, USA. Tel: 703 437 6191, 800 327 4536; fax: 703 435 9774; http://www.glenres.com.

(UK distributor, Cambio, 34 Newnham Road, Cambridge CB3 9EY. Tel: 01223 366 500; fax: 01223 350 069.)

Grant Instruments (Cambridge) Ltd, Barrington, Cambridge, CB2 5QZ, UK. Tel: 01763 260811; fax: 01763 262410. (USA distributor, Science Electronics.)

Hoefer Pharmacia Biotech Inc., 654 Minnesota Street, San Francisco, CA 94107, USA. Tel: 800 984 9947; fax: 415 821 1081.

Hybaid Ltd, 111–113 Waldegrave Road, Teddington, Middlesex, TW11 8LL, UK. Tel: 0181 614 1000; fax: 0181 977 0170. (USA distributor, National Labnet Co.)

INCSTAR Corporation, 1990 Industrial Blvd, Stillwater, MN 55082, USA. Tel: 612 439 9710; fax: 612 779 7847.

Incstar Ltd, Charles House, Toutley Road, Wokingham, Berks, RG41 1QN, UK. Tel: 0118 936 4200; fax: 01189 792 061.

Invitrogen Corporation, 3985-B Sorrento Valley Boulevard, San Diego, CA 92121, USA. Tel: 800 655 8288, 619 597 6200; fax: 619 597 6201. (UK distributor, R&D Systems Europe.)

Life Technologies Inc., 8400 Hergerman Court, Gaithersburg, MD 20884, USA. Tel: 301 840 8000; fax: 301 670 8539.

Life Technologies Ltd., European Division, PO Box 35, 3 Fountain Drive, Inchinnan Business Park, Paisley, PA5 9RF, UK. Tel: 0141 814 6100; fax: 0141 814 6258.

M.J. Research Inc., 24 Bridge Street, Watertown, MA 02172, USA. Tel: 800 729 2165; fax: 617 924 2148. (UK distributor, Genetic Research Instrumentation Ltd.)

NBL Gene Sciences Ltd, South Nelson Industrial Estate, Cramlington, Northumberland, NE23 9WF, UK. Tel: 01670 732 992; fax: 01670 730 454.

New England Biolabs Inc., 32 Tozer Road, Beverley, MA 01915-5599, USA. Tel: 508 927 5054; fax: 508 921 1350.

New England Biolabs, 67 Knowl Piece, Wilbury Way, Hitchin, Herts, SG4 0TY, UK. Tel: 01462 420 616; fax: 01462 421 057.

Novagen Inc., 601 Science Drive, Madison, WI 53711, USA. Tel: 608 238 6110; fax: 608 238 1388. (UK distributor, NBL Gene Sciences Ltd.)

Operon Technologies, Inc., 1000 Atlantic Ave., Suite 108, Alameda, CA 94501, USA. Tel: 800 688 2248, 510 865 8644; fax: 510 865 5255; email: dna@operon.com.

PanVera Corp., 545 Science Drive, Madison, WI 53711, USA. Tel: 608 233 9450; fax: 608 233 3007; email: info@panvera.com.

Perkin-Elmer, *see* Applied Biosystems.

Pharmacia Biotech (Biochrom) Ltd, 22 Cambridge Science Park, Milton Road, Cambridge, CB4 4FJ, UK. Tel: 01223 423 723; fax: 01223 420 164.

5 Prime → 3 Prime Inc., 5603 Arapahoe Road, Boulder, CO 80303, USA. Tel: 800 533 5703; fax: 303 440 0835. (UK distributor, CP Laboratories.)

Promega Corp., 2800 Woods Hollow Road, Madison, WI 53711–5399, USA. Tel: 800 356 9526, 608 274 4330; fax: 608 277 2516.

Promega Ltd, Delta House, Enterprise Road, Chilworth Research Centre, Southampton, SO16 7NS, UK. Tel: 0800 760 225, 01703 760 225; fax: 01703 767 014.

R&D Systems Europe Ltd, 4–10 The Quadrant, Barton Lane, Abingdon, Oxon, OX14 3YS, UK. Tel: 01235 551 100; fax: 01235 533 420.

Roche Diagnostic Systems, 1080 Highway 202, Bldg 500, Branchburg Township, NJ 08876-3771, USA. Tel: 908 253 7200.

Roche Molecular Systems, *see* Applied Biosystems.

Sanyo Gallenkamp plc, Park House, Meridian Business Park, Leicester, LE3 2UZ, UK. Tel: 0116 263 0530; fax: 0116 263 0353.

Sorin Biomedica S.p.a., via Crescentino, 13040 Saluggia (Vercelli), Italy. Tel: 161 487373; fax: 161 487642. (USA distributor, Incstar Corp. UK distributor, Incstar Ltd.)

Stratagene, 11011 North Torrey Pines Road, La Jolla, CA 92037, USA. Tel: 800 424 5444; fax: 619 535 5430.

Stratagene Ltd, Cambridge Innovation Centre, 140 Cambridge Science Park, Milton Road, Cambridge, CB4 4GF, UK. Tel: 0800 585 370, 01223 420 955; fax: 01223 420 234.

Stratech Scientific Ltd, 61–63 Dudley Street, Luton, Beds, LU2 0NP, UK. Tel: 01582 481 884; fax: 01582 481 895.

Takara Shuzo Co., Ltd, Biomedical Group, Seta 3-4-1, Otsu, Shiga 520-21, Japan. Tel: 0775 43 7235; fax: 0775 43 2312. (UK distributor, Stratech Scientific Ltd; USA distributor, PanVera Corp.)

Techgen International Ltd, Suite 8, 50 Sullivan Road, London SW6 3DX, UK. Tel: 0171 371 5922; fax: 0171 371 0496.

Techne Inc., 1743 Alexander Road, Princeton, NJ 08540, USA. Tel: 609 452 9275; fax: 609 987 8177.

Techne (Cambridge) Ltd, Duxford, Cambridge, CB2 4PZ, UK. Tel: 01223 832 401; fax: 01223 836 838.

Appendix C

Further reading

The following texts are recommended to those readers wishing to investigate PCR and the applications and properties of PCR-related techniques further.

Davies, K. (ed.) (1988) *Genome Analysis: A Practical Approach*. Oxford University Press, Oxford.

Dieffenbach, C.W. and Dveksler, G.S. (eds) (1995) *PCR Primer: A Laboratory Manual*. Cold Spring Harbor Laboratory Press, Cold Spring Harbor, New York.

Ellingboe, J. and Gyllensten, U.B. (eds) (1992) *The PCR Technique: DNA Sequencing*. Eaton Publishing, 154 E. Central Street, Natick, MA 01760, USA.

Erlich, H.A. (ed.) (1989) *PCR Technology*. Stockton Press, New York.

Griffin, H.G. and Griffin, A.M. (eds) (1994) *PCR Technology: Current Innovations*. CRC Press, Florida.

Innis, M.A., Gelfand, D.H., Sninsky, J.J. and White, T.J. (eds) (1990) *PCR Protocols: A Guide to Methods and Applications*. Academic Press, San Diego.

Landegren, U. (ed.) (1996) *Laboratory Protocols for Mutation Detection*. Oxford University Press, Oxford.

Larrick, J.W. and Siebert, P.D. (eds) (1995) *Reverse Transcriptase PCR*. Ellis Horwood Limited, Hertfordshire.

McPherson, M.J., Quirke, P. and Taylor, G.R. (eds) (1991) *PCR: A Practical Approach*. Oxford University Press, Oxford.

McPherson, M.J., Hames, B.D. and Taylor, G.R. (eds) (1995) *PCR-II: A Practical Approach*. Oxford University Press, Oxford.

Mullis, K.B., Ferry, F. and Gibbs, R.A. (eds) (1994) *The Polymerase Chain Reaction*. Boston.

Newton, C.R. (ed.) (1995) *PCR: Essential Data*. John Wiley and Sons, West Sussex.

Rolfs, A., Schuller, I., Finckh, U. and Weber-Rolfs, I. (1992) *PCR: Clinical Diagnostics and Research.* Springer-Verlag, Berlin.

White, B.A. (ed.) (1993) *Methods in Molecular Biology,* Vol. 15 *PCR Protocols; Current Methods and Applications.* Humana Press, New Jersey.

Amplifications; A forum for PCR users. A news sheet (free subscription) produced by P.E. Applied Biosystems.

BioTechniques, Eaton Publishing.

Genome Research, Cold Spring Harbor Laboratory Press. (This journal now incorporates *PCR Methods and Applications.*)

Nucleic Acids Research, Oxford University Press.

PCR Methods and Applications, Cold Spring Harbor Laboratory Press. A journal devoted solely to PCR, covering reviews, research reports and technical tips. Also includes PCR product advertisements from molecular biology companies and instrument suppliers. Published quarterly. (*See Genome Research.*)

Strategies. A quarterly newsletter produced by Stratagene.

INDEX